The Construction of New Mathematical Knowledge in Classroom Interaction

Mathematics Education Library

VOLUME 38

The titles published in this series are listed at the end of this volume.

Heinz Steinbring
(Author)

The Construction of New Mathematical Knowledge in Classroom Interaction

An Epistemological Perspective

Springer

Heinz Steinbring, Universität Duisburg-Essen, Essen Germany

Library of Congress Cataloging-in-Publication Data

Steinbring, Heinz.
The construction of new mathematical knowledge in classroom interaction: an epistemological perspective / Heinz Steinbring.
p. cm.—(Mathematics education library; v. 38)
Includes bibliographical references and indexes.
ISBN 0-387-24251-1 (acid-free paper) -- ISBN 0-387-24253-8 (E-Book)
1. Mathematics—Study and teaching. 2. Teacher-student relationships. 3. Interaction analysis in education. I. Title. II. Series.

QA11.2.S84 2005
510'.71—dc22 2004061914

Printed in the United States of America.

9 8 7 6 5 4 3 2 1 SPIN 11347412

springeronline.com

Dedication

This book is dedicated to
Christel, Kristin and Hanna

TABLE OF CONTENTS

ACKNOWLEDGEMENTS

First of all I am especially grateful to Anna Sierpinska for her professional support in organising and accompanying the editorial work of the book. On the basis of the work of her student, Varda Levy, who polished and edited the English language as a native speaker, Anna made corrections and suggestions for improvement from the point of view of a mathematics education researcher.

I have also to mention and to thank Kristin Steinbring, my daughter, for her help in preparing a first German-English version of the book.

Without the help of Kristin, Varda Levy and Anna Sierpinska the book would not have been produced in this form. Thank you very much.

PREFACE

Mathematics is generally considered as the only science where knowledge is uniform, universal, and free from contradictions. „Mathematics is a social product – a 'net of norms', as Wittgenstein writes. In contrast to other institutions – traffic rules, legal systems or table manners –, which are often internally contradictory and are hardly ever unrestrictedly accepted, mathematics is distinguished by coherence and consensus. Although mathematics is presumably the discipline, which is the most differentiated internally, the corpus of mathematical knowledge constitutes a coherent whole. The consistency of mathematics cannot be proved, yet, so far, no contradictions were found that would question the uniformity of mathematics“ (Heintz, 2000, p. 11).

The coherence of mathematical knowledge is closely related to the kind of professional communication that research mathematicians hold about mathematical knowledge. In an extensive study, Bettina Heintz (Heintz 2000) proposed that the historical development of formal mathematical proof was, in fact, a means of establishing a communicable „code of conduct“ which helped mathematicians make themselves understood in relation to the truth of mathematical statements in a co-ordinated and unequivocal way.

If this is how one thinks about mathematical knowledge and its communication then the question for mathematics education is if this philosophical position determines some definite mental model for the teaching and learning of mathematics. Are the processes of communication in mathematical instruction comparable to those held amongst professional mathematicians? Can they or should they even conform to the experts' forms of argumentation?

In this book, I take the following approach to this problem. First of all, I assume that students in mathematics instruction are only „on their way“ to becoming persons who communicate and argue mathematically. For a professional mathematician, the content related mathematical ways of argumentation are forms of communication that have become familiar through long experience. For children learning mathematics, however, mathematical argumentation represents a central content of their learning. Furthermore, the mathematical argumentation of young students is constrained by the epistemological conditions of mathematical knowledge. For students who are only beginning to learn mathematics, this knowledge is not accessible in the same abstract form as it is for professional mathematicians. It is situated and bound to concrete experiences; Bauersfeld speaks here of „subjective domains of experience“ (Bauersfeld, 1983).

A second, essential aspect that has to be emphasized is that, unlike in professional mathematical communication, communication between teacher and students in the context of mathematical instruction is determined by the intention of mediating and learning mathematical knowledge. The influence of instructional goals must be particularly taken into consideration in understanding and analyzing the specificity of interactive mathematical teaching and learning processes.

In this book I present the basic concepts of the theoretical framework and I outline the chosen research methodology, which I then illustrate by a range of empirical case studies of elementary school mathematics instruction, where I analyze the conditions of the interactive construction of new mathematical knowledge. This research points at two central dimensions in the interactive knowledge construction: (1) the (general) *communicative* dimension, and (2) the *epistemological* dimension of mathematical knowledge. Interactive constructions of new mathematical knowledge as well as the necessary generalizing justifications cannot be carried out by elementary school students with the „classical" mathematical concepts of elementary algebra. This means that children can rely neither on a general algebraic notation nor on the rules of algebra in describing the yet unfamiliar knowledge or in operating with it. In the frame of elementary school mathematics instruction, new knowledge, in its interactive development, is characteristically bound to the students' situated learning and experience contexts. Children have to learn – and are able to do so by their own means – to see the general (the new knowledge) in the particular.

The central concern of this book is to investigate more closely the particularities and the variety of children's interactive constructions and justifications of mathematical knowledge in everyday mathematics teaching. The analysis of communicative factors in the culture of mathematical instruction is built on a broad theoretical basis, where the epistemological conditions of mathematical knowledge are particularly related to interactive constructions of knowledge. This is completed by detailed, extensive case studies of teaching and learning processes in which mathematical knowledge is constructed. These cases are analyzed with an epistemology-oriented methodology and the outcome of these analyses shows the complexity of the forms of constructing and justifying mathematical knowledge in instructional interactions. They provide productive possibilities, but also point to the restricting constraints of these communications. The differences between the mathematical communication of students who learn and the unequivocal communications of professional mathematicians become clear in a very evident way. However, in spite of all differences and the missing (but also not to be expected) formal, abstract argumentation, it can be observed in many teaching episodes that children participate in true mathematical communications, especially when one takes into account the particular conditions of the situatedness of the mathematical knowledge and the influences of the instructional intention within the communication.

The present book is based on a foundational theory in mathematics education about the epistemology and learning of mathematical knowledge, developed over many years of theoretical work and discussion, especially with colleagues and friends at the Institute for Didactics of Mathematics at the University of Bielefeld. Furthermore, the empirical results reported in the book are gathered in a long-term research project on mathematics teaching and learning in elementary school. This research aimed at examining the use of the substantial mathematical learning environments (from the „mathe 2000" project) in the everyday mathematics classroom. The research project was funded by the German Research Community (Deutsche Forschungsgemeinschaft, DFG; topic: „Epistemological and Socio-Interactive

Conditions of the Construction of Mathematical Knowledge Structures (in Primary Teaching)", 1.4. 1997 until 31.3.2000; reference number: STE 491/5-1 & 5-2; cf. Steinbring, 2000).

At this point, I would like to thank the participating teachers for agreeing to actively co-operate in the research project, besides having to fulfill their everyday teaching duties. I thank them for preparing the partly unfamiliar new teaching units, for carrying them out it their classes and for letting their instruction be audio- and videotaped. Without these teachers' willingness and engagement, this research project would not have been possible.

GENERAL OVERVIEW OF THE BOOK

This book deals with the following central problem in mathematics education: An essential concern of mathematics teaching consists in the requirements that the teacher is to develop, in the interaction with his or her students, mathematical understanding, competencies of explaining and reasoning as well as new mathematical knowledge. Briefly, in mathematics teaching, the students are supposed to learn mathematics. This leads to the question: How can everyday mathematics teaching be – as properly as possible – described, understood and developed as a teaching and learning environment in which the students gain mathematical insights and increasing mathematical competence by means of the teacher's initiatives, offers and challenges? How can the „quality“ of mathematics teaching be realized and appropriately described? Within this spectrum of rather general questions the following more specific research question is investigated in this book: How is new mathematical knowledge interactively constructed in a typical instructional communication among students together with the teacher?

Such a fundamental and far-reaching question has been the subject of investigation of many studies in mathematics education, using quite different approaches and methods. In order to deal with this question practice-oriented approaches for example would use the construction of good *classroom material* and the development of practical offers for the pedagogical designing of the instruction as well as the elaboration and trial of teaching methods for supporting the students' learning processes. In other research to analyze the question of what effective mathematics instruction could look like, for example, the students' learning successes are assessed and from these, the quality of the respective mathematics instruction is inferred.

Any research which aims at improving mathematics teaching by means of developing classroom material and teaching methods or by means of assessing the students' actual learning level and performance is based – deliberately and explicitly or unintentionally and subconsciously – on assumptions of the most diverse kind. These assumptions concern the particular role of mathematical knowledge, assumptions about ways of understanding mathematics, about how teaching and also learning works, and especially about what makes „good mathematics teaching“.

Those taking part in research and development of mathematics teaching have often been active for a long time within the milieu of teaching and learning mathematics in schools and high-schools. Consequently, in their scientific community, they have – partly explicitly, partly implicitly – adopted developed norms, standards and views. Among others, these are assumptions regarded as undisputed, about mathematics instruction, teaching and learning mathematics and about good or bad teaching, as they are widely spread and generally accepted as valid. Meanwhile, however, such self-evident assumptions, which have been partly easily accepted by some researchers and by many mathematics teachers several years ago, are being questioned more and more.

From a more external point of view, mathematics teaching is understood – at least by many teachers in their everyday practice – as an event in which the knowing teacher prepares the mathematical subject matter and then gives it to the students in appropriate portions. According to this view, mathematics is a ready made subject, which can be refined for the students' learning, but finally has to be accepted in the way it was produced by mathematicians.

This book, on the other hand, proceeds from the assumption that mathematics teaching with its intentions and obligations represents a complex structure and that it is necessary to enter more deeply under the visible surface and to question conventional assumptions more carefully. In order to meet this demand, the following two central levels shall be put in a reciprocal relation in the course of this book:

- the careful and detailed analysis of selected episodes out of everyday mathematics teaching by means of using an elaborated qualitative analysis method
- the explicit development of fundamental, theoretical views about the content of instruction, mathematical knowledge, interactive development of mathematical meanings within the course of teaching and about the conditions and possibilities of communication processes within – especially mathematical – interactions between teacher and students.

This complementary bracket of theoretical foundation and empirical analysis has the following perspective in view. Mathematical teaching processes are autonomous and stable processes, which have to be taken seriously as independent cultural events. This means that one has to, in a manner of speaking like an ethnographer, take part in the daily events occurring in the culture of mathematical teaching, perceive them carefully and want to understand the variety of the complex details. Here it is important not to have already made hasty valuations on the basis of unreflected assumptions about whether the mathematics instruction was successful or not. Only a sensitive and careful analysis of the instruction events can subsequently make it possible to conclude differentiated estimations about essential qualities of mathematics teaching.

This elaboration of fundamental, theoretical views about the epistemological nature of mathematical knowledge, about the interactive development of mathematical meanings within the course of instruction and about the conditions and possibilities of communication in mathematical teaching is presented in the first chapter of the book.

This more precise characterization of the theoretical views about mathematical knowledge, about the interactive development of mathematical meaning and about (mathematical) communication processes is a necessary foundation for access to the empirical phenomena of the complex events within mathematics teaching. The theoretical foundation and the empirical analysis need each other. In order to go deeper under the visible surface of the observed events in mathematics teaching, „theoretical glasses“ are required, which allow for recognizing important differences and particularities. Then the relational structure of the particularities has to be examined more closely with a „theoretical magnifying glass“. Finally, a „theoretical microscope“ is necessary in order to unearth connections and relationships deeper under the surface, which are supposed to contribute to understanding the observed mathematical interaction events in their manifold complexity.

With the sequence „glasses“, „magnifying glass“ and „microscope“, the development and refinement of the instruments for the use of the epistemologically oriented qualitative instruction analysis are to be illustrated. This development of the analysis instrument is based on the one hand on methods of qualitative researches about mathematics teaching and also on my own research work about the epistemology of mathematical knowledge. On the other hand, in this book (in chapter 2) the analysis instrument specifically for the investigations undertaken on the essential relationships between the communicative and epistemological conditions for understanding interactions in mathematical teaching is further elaborated. In particular the explication of the view about the epistemological character of mathematical knowledge is a foundation for this. In the center stands the mathematical sign or symbol, which on the one hand plays a central role for the epistemological particularity of mathematical knowledge, and which on the other hand acts as carrier of mathematical knowledge within mathematical communication and is used by the participants in the communication in order to talk about mathematics.

The so-called „epistemological triangle“ (view chapter 1, section 1.1) represents a conceptual scheme in order to characterize the epistemological particularity of mathematical knowledge, in which a distinctive reference is made to the role of mathematical signs or symbols. Besides the use of the epistemological triangle in order to clarify fundamental assumptions about mathematical knowledge, this epistemological triangle serves at the same time as an instrument of an (epistemologically oriented) analysis of episodes taken from everyday mathematics teaching.

The further development of the analysis instrument carried out in this book takes place by means of a conceptual connection of the epistemological triangle with a communicative analysis of the instruction interaction. This communicative analysis uses the fundamental elements of the „Theory of Society“ by the German sociologist Niklas Luhmann, who refers to the basic differentiation of sign in signifier and signified when characterizing the concept „communication“. With this, essential foundations for the analysis instrument (theoretical magnifying glass) are presented.

As any teaching, mathematics teaching is an interactive process, in which it is about developing new knowledge. The view about what new mathematical knowledge is can consist simply in the fact that in comparison to the knowledge that one already knows and that has already been treated in class, the unfamiliar and not yet treated knowledge represents the *new* knowledge. This view corresponds very much to the idea that mathematical knowledge is a finished and given product to which further knowledge elements can simply be added. This view is fundamentally questioned here. New mathematical knowledge is not merely still unfamiliar, added finished knowledge, but new mathematical knowledge has ultimately to be understood as an extension of the old knowledge by means of new, extensive relations, which at the same time let the old knowledge shine in a new light and, even generalize the old knowledge. This important view about the relation between old and new mathematical knowledge is elaborated in chapter 2, section 2.1.

Thus all the important components of the theoretical foundation are put together in order to carefully analyze in this research project such interactive mathematical teaching processes, which aim at the development of new mathematical knowledge. The methodical procedure for the qualitative analysis of selected instruction episodes is carried out in three steps: (1) „description of the episode along single phases“ (theoretical glasses); (2) „general epistemologically oriented qualitative analysis (theoretical magnifying glass); (3) „detailed analysis of interactive mathematical reasoning out of a communicative and epistemological view“ (theoretical microscope; cf. to this sections 1.2 and 3.2). The „result“ of this extensive analysis of mathematical interactions is classified in an analysis grid (developed in section 2.3).

This methodical analysis grid presents the two dimensions „epistemological characterization of the mathematical knowledge“ and „communicative interpretation of the mathematical knowledge“ into a relation with each other. The scope of these dimensions reflects the problem of the relationship between „old and new mathematical knowledge“ in a correspondingly specific way. Along the epistemological dimension, the interest is in the relation from an empirical towards a relational knowledge conception; and in the communicative dimension, the communication of fact knowledge is contrasted to the communication of relational mathematical meaning constructions (see to this section 2.3).

The elaboration of the theoretical foundations is thus finished within the first two chapters; at the same time, there is a necessary important anticipation of the method and content orientation of the empirical qualitative analysis, especially as far as the theoretical perspective is concerned, towards which a thorough access to the empirical phenomena is aimed.

The extensive and careful analyses of ten selected teaching episodes are presented in the central chapter 3. The episodes originate from elementary school mathematics instruction (3rd or 4th grade). In the observed and videographed instruction, problem-oriented mathematical learning environments were treated. The particular character of these mathematical exercises is described in section 3.1. With these exercises, challenging offers and additional suggestions for the interactive production of new mathematical knowledge, the central concern of any mathematics instruction, are given.

In section 3.3, the extensive qualitative instruction analyses are elaborated. These analyses run through the three mentioned methodical steps. On the basis of the previously elaborated theoretical foundations, the completed analyses make it possible to trace and to better understand the variety, independence and the respective particularities of the mathematical interactions and knowledge constructions of the students with their teacher in the course of instruction.

The particular epistemological difficulty of mathematical knowledge – contained in the specific role of the mathematical signs and symbols – consists in the fact that mathematical knowledge does not simply relate to given objects, but also that relations, structures and patterns are expressed in it. This relational character of mathematical knowledge is described and analyzed with the help of the epistemological triangle. Because of the relational character of mathematical knowledge, the students are faced with a particular challenge in their processes of learning, understanding and making their independent constructions of mathematics.

Learning and understanding – also of mathematics – is strongly bound to the senses and thus to perceivable phenomena and objects. Mathematical objects, however, are ultimately not directly visible or perceivable; they are ideal structures. The beginning learning process of the children starts from given, concrete and illustrative material as embodiments of mathematical structures (in arithmetic or geometry) and the requirement is to more and more understand and interpret these given objects as carriers of structures and relations. In this way, the specific epistemological character of mathematical knowledge in learning and construction processes can be accentuated.

The difficulty of understanding mathematical knowledge as relational knowledge and not as fact knowledge about concrete objects is a particular challenge for learning, understanding and construction processes. The analyzed episodes illustrate the problem of potentially interpreting the given concrete objects and reference contexts for mathematical signs and symbols in the teaching interaction as relational structures. This challenge does not always succeed; some interactions remain on the level of a concrete interpretation of the mathematical knowledge as fact knowledge. However, there is a number of episodes in which – of course in a way which is bound to the concrete situation and to the mathematical exercise examples – the requested interpretation of the reference objects as carriers of mathematical relations succeeds quite well.

In chapter 4, a classification of the analyzed episodes with the interaction events is performed using tan elaborated analysis grid. A comparative classification of the ten episodes shows the existing different interactive ways of interpreting mathematical knowledge (in the two described dimensions). For this, each of the analyzed mathematical meanings and constructions is assigned to one of the nine boxes in the analysis grid.

On the basis of the classification of the analyzed teaching episodes, three forms of interactive mathematical knowledge constructions are then (in section 4.4) described: (1) „knowledge construction as continuation of familiar, situated fact knowledge", (2) „knowledge construction as balance between consistent base knowledge and new knowledge relations" and (3) „knowledge construction as an introduction of isolated knowledge structures". For mathematics instruction, the second type of knowledge constructions is interpreted as the comparatively most productive kind of interactive production of new mathematical knowledge. In these interactive constructions, the problem of the relation between old and new mathematical knowledge is taken up in an appropriate way. All the analyzed mathematical interactions with the knowledge constructions are assigned to these three types.

An essential result of the book is the following: Opposed to an assessment of interactive events in mathematics instruction which rather remains at the surface of observing external phenomena, a careful and extensive qualitative analysis, which concentrates on epistemological and communicative problems, can bring multiple details, connections and interpretations in the interactive constructions of mathematical knowledge to the surface which otherwise remain hidden and undiscovered. In order to do this, it is necessary to replace the implicit and seemingly self-evident assumptions about mathematics teaching by the explication of differentiated theoretical positions. On this basis only, the indispensably precise instrument of analysis can be developed.

The important theoretical distinction between old and new mathematical knowledge and the epistemological particularity that mathematical signs and symbols do not simply relate to concrete things, but that only through them relations and structures are expressed, are fundamental conditions for the functioning of mathematical communication and construction processes. The detailed analyses of the instruction episodes give evidence that indeed such true mathematical communication and construction processes, that depend on many difficult conditions, can take place in mathematics instruction (even in elementary school). Furthermore, these analyses point out *in which way* these communications in mathematics teaching happen under the particular conditions of the actual learning situation and the children's interpretations and constructions.

The construction of mathematical knowledge seen as a balance between a consistent base knowledge and new knowledge relations is essential in order to find useful answers to the question asked at the beginning: How can everyday mathematics teaching be – as properly as possible – described, understood and developed as a teaching and learning environment in which the students gain mathematical insights and increasing mathematical competence by means of the teacher's initiatives, offers and challenges? How can the „quality" of mathematics teaching be realized and appropriately described?

OVERVIEW OF THE FIRST CHAPTER

Mathematical knowledge is the specific subject matter of the teaching and learning processes in mathematics instruction. At first, this is a mere truism. A problem that is worth thinking about emerges when one asks about the character or the epistemological nature of mathematical knowledge. A common answer is that mathematics represents an objective and logically consistent body of knowledge, which is produced or discovered in reality according to eternal laws and ideal objects by mathematical researchers. Such a view understands mathematics as an autonomous, already existing and ideal object, and any effective influence from research or also learners on this ideal mathematics is negated.

In this book, on the other hand, a different view onto the nature of mathematical knowledge is taken explicitly. The emerging of mathematical knowledge is fundamentally taking place in the context of social construction and individual interpretation processes. Mathematical knowledge is thus not previously a given; it is constructed by means of social activities and individual interpretations.

The view of the nature of mathematical knowledge, which is taken in this book, consequently always connects mathematical knowledge with the social context of research or of learning. In the elaboration of this fundamental position (section 1.1), the following problems will be discussed. The socio-historical development of mathematics has led to a particular uniformity and universality of mathematical knowledge as compared to other sciences. This uniformity, however, is not to be misunderstood as the expression of a supposed absolute objectivity of mathematics. This uniformity is the result of the professional work of researching and communicating mathematicians, thus emerging in from professional activity and was not simply objectively available beforehand.

Opposite to the mathematical research practice, the practice of teaching and learning mathematics is characterized by a variety of mathematical constructions and interpretations. By means of the instruction, the learning students are supposed to slowly become active members of the mathematical community; they are introduced to the mathematical culture. The coherent professional mathematical communication, which can be observed in particular in the reciprocal clarification of mathematical proofs, and which finally makes up the uniformity of mathematical knowledge, is not yet a self-evident basis of instructional communication for learning students, but a long-term goal.

In order to clarify the epistemological nature of mathematical knowledge, mathematics is always regarded in the context of an accompanying culture in this book (cf. section 1.1). The mathematical knowledge which is historically developed and handed over from one to the next generation belongs in the socio-historical culture. The academic knowledge emerges in the culture of the researching professional mathematicians, and finally, the up-and-coming generation of a society acquires the relevant mathematical knowledge in the teaching culture. Within a culture, especially the respective signs and symbols as well as their use and interpretation play a central role.

The mathematical signs and symbols have an outstanding meaning in the particular mathematical culture. During the long development of the socio-historical culture, the emergence of signs and symbols as well as changes in their use and their interpretations can be observed. In the professional culture, mathematical signs are used by the participants in the communication in a more or less unambiguous and well-defined way. In the teaching culture, the students have to be introduced into the use of these mathematical signs and symbols, and therefore a variety and a certain diversity of the mathematical communication in teaching and learning processes can be observed.

Mathematical signs and symbols have a double function for communicative processes of teaching and learning mathematics. On the one hand, they are the „carriers" of mathematical knowledge; i. e. with their help mathematics is written down and represented. On the other hand, the mathematical signs in different forms and ways of use and interpretation are the central elements of communication within the instruction culture. This makes it clear that research problems in mathematics education, which aim at real teaching processes, their careful description and thorough analysis, must ascribe an prominent position to the mathematical signs / symbols.

The epistemological triangle is used as a central theoretical concept in this book in order to adequately describe the epistemological particularity of mathematical knowledge (especially within the instruction culture), and in order to analyze mathematical communication processes within instruction. In the frame of the epistemological triangle, the (double) function of the mathematical sign / symbol is characterized (section 1.1). The mathematical signs obtain their meaning only by means of a relation to reference contexts; ultimately, the learning students must actively produce this relation themselves. With the help of examples taken from (elementary) school mathematics, the functioning of the epistemological triangle will be explained.

Furthermore, it will be made clear what the particular character in the reference relation between the mathematical signs / symbols and the respective reference context consists of for mathematical knowledge. For instance, this is not about a simple relation in which the sign is linked as a name with an existing object in reality. The core of this relation consists in the fact that ultimately, signs / symbols and reference contexts must be seen as structures full of relations between whom a mediation takes place. Additionally, this mediation is steered and regulated by means of fundamental mathematical conceptual aspects. These epistemological particularities of mathematical signs or symbols are concretely explained with the help of mathematical examples (among others the probability concept).

In section 1.2, the theoretical foundations and methods of the epistemologically oriented mathematical instruction research are represented. In the frame of the qualitative interpretative research approaches to mathematics teaching, the epistemologically oriented interaction research focuses in its analysis on the interactive development of mathematical meanings and knowledge constructions. In this research paradigm, especially the epistemological problems of the production of appropriate mediations between mathematical signs / symbols and matching reference contexts are analyzed, regarding school mathematics as an existing knowledge domain and paying attention to mathematical interaction processes of the construction of mathematical knowledge.

With reference to familiar essential works of the epistemologically oriented instruction research, methodical steps in the realization of the epistemological analysis are represented and carried out at an instruction example (section 1.2). The epistemological triangle and its application will be explained one more time during the qualitative analysis of this example.

Mathematical teaching communications usually take place in the frame of general communication. Under the fundamental perspective taken here, according to which mathematical knowledge development is always embedded in accompanying social contexts, communication is particularly important, be it in the specific form of a mathematical communication, or in the extensive form of any communication. In section 1.3, the theoretical position of the sociologist Niklas Luhmann is taken in order to clarify the role of „communication“. According to Luhmann, communication is the essential element, which „defines“ from a sociologist perspective, what society means. Communication represents an (autopoietic) system. It must be distinguished from individual consciousness and from individual cognition.

In order to substantiate the autopoietic system „communication“, Luhmann refers to the distinction of sign into signifier and signified according to de Saussure. In section 1.3, the autonomous functioning of the communication system (according to Luhmann) is described and explained with a mathematical example.

Furthermore, the interpretation of the mathematical signs / symbols in the epistemological triangle is presented in a relation to the communicative interpretations of the signifiers in the frame of the (general) communication as an autopoietic system. This section makes it clear that on the one hand, the communication between teacher and students in the instruction can be seen as a component of everyday communication, and that on the other hand, the particular character of mathematical communication must be paid attention to.

CHAPTER 1

THEORETICAL BACKGROUND AND STARTING POINT

The focus of this research is the analysis of essential conditions for the social development and interactive constitution of (new) mathematical knowledge in everyday elementary teaching. On the one hand, the project fits into the mathematics-didactic research field of qualitative and quantitative studies on mathematics teaching. On the other hand, there are important connections to empirical sociological studies related to the conditions of the genesis of scientific – and especially mathematical – knowledge in social practices or institutions.

A basic assumption is that mathematics, like any other theoretical knowledge, always needs specific contexts in which it develops, organizes, becomes systematized, and connects to meanings. The nature of possible interrelations between a social developmental or practical context and the emergence of mathematical knowledge is examined as a research question for *scientific*, mathematical knowledge. For example, mathematical knowledge, or the science „mathematics“ can be understood as a socio-historical culture in its historical development (cf. e.g. Wilder, 1981). On the other hand, from a sociological perspective, current pure mathematics is characterized as universal knowledge which emerges amongst mathematicians through highly specialized and formalized communicative argumentative processes – creating socially negotiated, coherent proof (Heintz, 2000). A question is also raised as to the interrelationships between the genesis of knowledge and the institutional, social environment in the application of knowledge, for example, mathematical knowledge which is applicable to other domains of use (for a sociological analysis of the emergence of theoretical knowledge in laboratory practice see Knorr-Cetina, 1981).

When dealing with questions of the acquisition of mathematical knowledge in *the context of the classroom and other learning situations*, a very important role is given to the connection between social conditions, influencing factors, and the mathematical knowledge to be acquired and developed. Thus, in the didactics of mathematics, the mutual relation between the emergence of (school) mathematical knowledge and the accompanying institutional or common social environment and practices has been formulated as a central problem, examined from different theoretical viewpoints and treated with different conceptual approaches. As a fundamental concept, Bauersfeld (1983; 1994) for example, develops the concept of „domains of subjective experiences“, in which young children embody their mathematical ideas in concrete situations, notions, and connections, which they are able to fill with meaning derived from their social environment. The children can also develop and generalize their mathematical concepts, over partially conflicting domains of subjective experiences (concerning this, see also the concept of „micro worlds“ (Lawler, 1990) quoted by Bauersfeld). The research school of „situated cognition“ (Lave, 1988; Lave & Wenger, 1991; Wenger 1998) which, over the last years, has been broadly recognized and adopted by many researchers, exemplifies an important research approach which stresses the relevance of social conditions for individual construction of mathematical knowledge, both in and out of school.

In this work it is assumed that there exists an unalterable reciprocal interrelationship between mathematical knowledge and social contexts. This is indispensable for the scientific investigation of the interactive development of mathematical knowledge in the school classroom similar to other fields of mathematical practice. The interrelationship can be illustrated in the following diagram (Fig. 1).

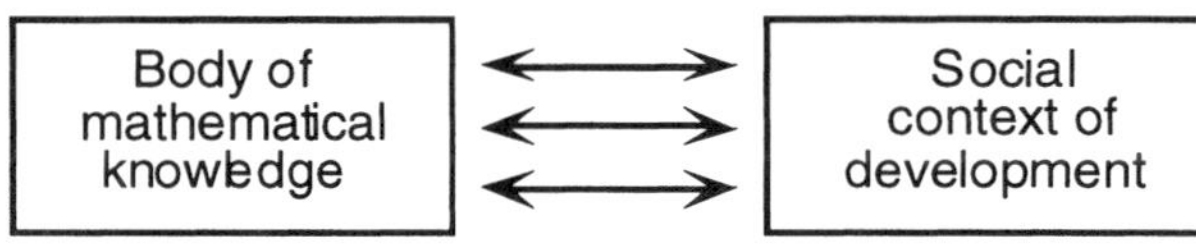

Figure 1. Interrelationship between knowledge and social contexts.

This interrelationship expresses the basic view that mathematical knowledge cannot be constructed independently from a social development context. However, this knowledge is not completely subject to the social conditions, but is, at the same time, bound to epistemological constraints and logical consistencies (cf. Indurkhya, 1994, p. 106).

From this perspective of analyzing the interactive development of mathematical knowledge in the social context of everyday teaching (for instance in elementary school), a series of important questions is immediately raised. The first group of questions refers to the mathematical knowledge. On which conception of the epistemological nature of mathematical knowledge does it make sense to proceed, in the aforementioned reciprocal interrelationship? The assumption of an exclusively objectivistic interpretation of mathematics as a completely independent and autonomous knowledge domain, often proclaimed, especially by mathematicians, on the basis of different philosophical positions, would negate any essential relation to a social environment. On the other hand, a complete reduction of the emergence of mathematical knowledge to the social context – e.g. the classroom situation – could not explain the „convergence“ with definite statements, connections, and structures in mathematical knowledge. Different social practices could eventually lead to diverging epistemological mathematical knowledge interpretations (concerning the scientific-sociological analysis of the universality of mathematics in its development in the social practices of the pure mathematician researcher, see especially the well-founded work by Heintz, 2000).

A second group of questions deals with the role of social contexts in the development of knowledge. The particularities of the accompanying social contexts – in this work, elementary mathematics teaching with interactions between learners and the teacher – represent particular influencing factors in the development of mathematical knowledge within the learning processes. What role, in the learning of mathematics, do the social development contexts play? How do these contexts differ from other professional social contexts? And, does mathematical knowledge, constituted in the context of social learning, fundamentally differ from knowledge created in research contexts. Could schoolmathematics therefore be a separate kind of mathematics, or would it only differ with regard to the degree of formalization in its terminological and argumentative notation? In other words, has school mathematics, in principle, the same epistemological status as scientific mathematics? The general approach to answering these questions will also demonstrate that scientific mathematics and school mathematics are similar with respect to their social contexts and their fundamental epistemological status, but that they differ considerably in regard to their degrees of formalization, their purposes in the learning of mathematics, and the different tasks of mathematics education.

> ... every mathematical knowledge, be it scientific knowledge or school knowledge, needs reference contexts, and, in its sense, every knowledge is context-specific. On this basis, the difference between scientific and school mathematics lies in the different types of reference contexts used in these different social contexts of development. One important difference concerns the reference contexts in school mathematics which must be adjusted to the requirements of learning and to the cognitive development of the students (Steinbring, 1998a, p. 524).

This comparison between scientific and school mathematics leads to the third group of questions. What is the specific interest in the analysis of the relation between social contexts and the development or acquisition of mathematics, especially from the perspective of mathematics education? On the one hand, several specific characteristics in terms of the social context of school teaching and classroom interaction, as opposed to the scientific development of mathematical knowledge will be identified. On the other hand, unlike mathematicians‘ research practice (cf. Heintz, 2000), one can expect a bigger difference and variability in the teaching and learning processes, which is to be analyzed and understood from certain perspectives. The evaluation of different social learning situations with regard to the epistemological content of the particular interactively produced mathematical knowledge is a genuine concern for mathematics education. A more profound understanding of the interaction mechanisms in mathematics teaching, that is based on careful empirical analyses and theoretically grounded concepts, is necessary for introducting a distinction between successful and unsuccessful or poor interactive classroom processes of constructing mathematical knowledge. Such an evaluation of mathematical learning situations could also offer constructive didactics of mathematics (that develops mathematical learning environments) new possibilities for deciding about changes and variations in the construction of the mathematical learning environments on the basis of the insights gained from qualitative analysis.

Against this background of a fundamental relation between the context of social development and the construction of mathematical knowledge, and in connection to the questions just touched upon, the particular epistemological status of mathematical knowledge will be elaborated both generally and by examples of teaching processes. An explanation of this particular interpretative research approach as an *epistemologically-oriented analysis* of mathematical classroom interactions, will follow, along with the description of selected works in this field. The characterization of communication as an autonomous system (according to Luhmann, 1997a) supports this epistemologically-oriented research setting and raises additional fundamental questions pertaining to this research work.

1. BETWEEN UNITY AND VARIETY – CONCEPTIONS OF THE EPISTEMOLOGICAL NATURE OF MATHEMATICAL KNOWLEDGE

Mathematics is usually considered as the science par excellence, in which there exist universal and definite results expressing indubitable truths. This concept of uniform mathematics implies, among other things, that no different „mathematics" or opposite viewpoints, nor contradictory or incompatible positions exist in mathematics. In this sense, mathematics is *the uniform* mathematics, in spite of the specialization and differentiation in its many subdisciplines. Bourbaki describes his thesis of uniform scientific mathematics in an exemplary and paradigmatic way. This Bourbakist viewpoint has also substantially influenced the characterization of the structure of school-mathematical knowledge. In his architecture of mathematics, Bourbaki writes:

> ... it is possible to ask whether this luxuriant proliferation [of mathematics] is the growth of a vigorously developing organism which gains more cohesion and unity from its daily growth, or whether on the contrary it is nothing but the external sign of a tendency toward more and more rapid crumbling due to the very nature of mathematics, and whether mathematics is not in the process of becoming a Tower of Babel of autonomous disciplines, isolated from one another both in their goals and in their methods, and even in their language. In a word, is the mathematics of today singular or plural? (Bourbaki, 1971, p. 24).

He answers the question himself, in that he works out the core of *uniform* mathematics by the axiomatic method and by types of structure.

A convergence to a uniform coherence of mathematical knowledge over all mathematical subdisciplines is an incontestable and striking feature of the academic mathematical discipline compared to other sciences (cf. concerning this point the book by B. Heintz, 2000). Presently, nothing is is being stated about the way in which the unity is produced and substantiated, whether because mathematical knowledge is seen as a priori knowledge – e.g. Platonic knowledge –, independent from any empiricism, or because this unity is the result of a socio-historical, interactive process between mathematicians. In her extensive work, Heintz (2000) develops a scientific-sociological interpretation of the fact that scientific communication between mathematicians does not lead to contradictory or even different points of view – as in other sciences – but that this communication has developed a very particular form, the formalized mathematical proof. Accordingly proof is not seen as an objectively valid instance of control completely independent from human beings. Instead, according to Heintz (2000), as a body of well-rehearsed rules, the proof guides the understanding in mathematicians' communication. Communication between mathematicians, as controlled by these rules, leads to the unity and coherence of mathematical knowledge.

As a consequence of the perceptible unity and coherence of scientific mathematics, mathematical knowledge as a whole, especially the mathematical content taught and learned at different school levels, is also regarded as uniform knowledge, equal and invariant for all intents and purposes. The image of uniform scientific research mathematics determines the idea of mathematical knowledge in its different developmental and applied fields: mathematical knowledge is generally conceived of as objective and therefore ready made and absolutely valid knowledge.

The unity of scientific mathematics is the result of a socio-historic and interactive communication process between mathematicians, which is, in a way, oriented towards a coherent product – that of uniform mathematics. Dealing with this, one has to distinguish between the development process and the target product. In scientific and widely differentiated mathematics research, the correct, universally valid mathematical (research) product obviously stands at the center in the dialectics of process and product. By contrast, other developmental and applied fields of mathematics may focus on other features in the dialectics of process and product. Here, the development *process* to a mathematical product often, by necessity, is central. The shift, from the mathematical product to the mathematical process, is an important issue in the learning and the active acquisition of mathematical knowledge, especially within the context of knowledge mediation in the classroom.

Freudenthal has emphasized the process character of mathematics for learning in a paradigmatic way.

> It is true that words as mathematics, language, and art have a double meaning. In the case of art it is obvious. There is a finished art studied by the historian of art, and there is an art exercised by the artist. It seems to be less obvious that it is the same with language; in fact linguists stress it and call it a discovery of de Saussure's. Every mathematician knows at least unconsciously that besides ready-made mathematics there exists mathematics as an activity. But this fact is almost never stressed, and non-mathematicians are not at all aware of it (Freudenthal, 1973, p. 114).

Mathematics, as an activity, implies that learning becomes an active process in the construction of knowledge.

> The opposite of ready-made mathematics is mathematics *in statu nascendi*. This is what Socrates taught. Today we urge that it be a real birth rather than a stylized one; the pupil himself should re-invent mathematics. ... The learning process has to include phases of directed invention, that is, of invention not in the objective but in the subjective sense, seen from the perspective of the student (Freudenthal, 1973, p. 118).

Apart from focusing on a finished, generally valid mathematical (research) product, development processes are neither uniform, universal, nor homogeneous. Subjective characteristics of those keeping the process going, as well as situated representations, notations and interpretations of mathematical knowledge, are manifold, divergent, and partly heterogeneous. In the process of developing mathematical knowledge, cultural contexts, subjective influences, and situated dependencies are both effective and inevitable, and are the reasons for an observable diversity and a non-uniformity of the emerging knowledge. In this regard, a learning student cannot be compared with a professional mathematician. The latter has many years of experience in mathematical communication with his colleagues, in the negotiation of the correctness of a mathematical assertion by using the communicative rules of a formal proof. Such professional communication aims directly at the uniform mathematical product in question, while the learning student is requested to develop and perfect such forms of mathematical communication with his classmates. The latter development process is essentially influenced by cultural aspects of teaching, by learning conditions which are subjective, by individual cognitive abilities, and by situated exemplary mathematical expressions and interpretations. Therefore, in the process of developing, learning and imparting mathematics, divergence and nonuniformity in understanding and interpreting are central.

The contrast between uniform scientific mathematics – oriented towards the generally valid (research) product – and the different perspectives and interpretations of mathematics produced in social environments for different application domains – tied up in situatedly framed development processes – becomes extremely apparent against the background of the different cultures in which mathematical knowledge is used and experienced. The culture of the researching and teaching mathematician, and the culture of mathematics teaching, face one another in an obviously distinct, and sometimes opposing, way. The role the Bourbakist mother structures play for the unity of mathematics cannot be understood by mere appropriation of the principles given by these structures. The culture of mathematical science and the historical development of mathematics form the necessary background for an understanding. These principles, of the unity of mathematical knowledge, cannot easily be transferred to school mathematics. With such an endeavor, school mathematics would lose their cultural background and become mere formalistic signs and formulas. In order to understand these signs and formulas, again, the formation of a new, distinct culture would be necessary, a kind of mathematical re-invention (Freudenthal). From the point of view, that mathematical signs, symbols, principles, and structures can only be meaningfully interpreted in the frame of a grown or newly emerging culture, one has to question the unity of mathematics in learning and teaching processes. If mathematical knowledge (signs, symbols, principles, structures etc.) can only be meaningfully interpreted in the frame of a specific cultural environment, then there is not simply one single, but many different forms of mathematics.

Wittmann discusses the problem of uniformity and difference of mathematics, not from the view of mathematical science like Bourbaki, but from the fundamental perspective of different scientific and practical fields, of society and culture. Wittmann distinguishes between specialized, scientific mathematics and the general social „phenomenon" mathematics.

> [One] ... must conceive of ‚mathematics' as a broad societal phenomenon whose diversity of uses and modes of expression is only a part reflected by specialized mathematics as typically found in university departments of mathematics. I suggest a use of capital letters to describe MATHEMATICS as mathematical work in the broadest sense; this includes mathematics developed and used in science, engineering, economics, computer science, statistics, industry, commerce, craft, art, daily life, and so forth according to the customs and requirements specific to these contexts. Specialized mathematics is certainly an essential element of MATHEMATICS, and the broader interpretation cannot prosper without the work done by these specialists. However, the converse is equally true: Specialized mathematics owes a great deal of its ideas and dynamics to broader scientific and societal sources. By no means can it claim a monopoly for ‚mathematics'.
>
> It should go without saying that MATHEMATICS, not specialized mathematics, forms the appropriate field of reference for mathematics education. In particular, the design of teaching units, coherent sets of teaching units and curricula has to be rooted in MATHEMATICS.
>
> As a consequence, mathematics educators need a lively interaction with MATHEMATICS, and they must devote an essential part of their professional life to stimulating, observing, and analyzing genuine MATHEMATICAL activities of children, students and student teachers. Organizing and observing the fascinating encounter of human beings with MATHEMATICS is the very heart of didactic expertise and forms a natural context for professional exchange with teachers.

> As a part of MATHEMATICS, specialized mathematics must be taken seriously by mathematics educators as one point of view that, however, has to be balanced with other points of view (Wittmann, 1995, p. 358/9).

With regard to the question of the unity or the variety of mathematical knowledge the following statement can be made. One should not naively proceed either from a complete unity or from an arbitrary variety of mathematical knowledge. The problem lies deeper: Mathematical knowledge cannot be revealed by a mere reading of mathematical signs, symbols, and principles. They have to be interpreted, and this interpretation requires experiences and implicit knowledge – one cannot understand these signs without any presuppositions. Such implicit knowledge, as well as attitudes and ways of using mathematical knowledge, are essential within a culture. Therefore, the learning and understanding of mathematics requires a cultural environment. The professional mathematician has been introduced to the culture of his mathematical research domain long ago and in his career he has acquired the implicit knowledge necessary for organizing his research work and communicating with colleagues. By contrast, a central problem for the learner of mathematics in school is of just being introduced to an appropriate mathematical culture and being able to participate in it.

With this distinction, of mathematical knowledge between „*mathematical signs, symbols*“ and a „*cultural environment*“ belonging to it, the question of „unity vs. variety“ can be examined more closely. On the side of the signs – more precisely with respect to the mathematical structural relations represented by the mathematical signs – there can exist a relative unity of mathematics (in some parts), and there is also a sort of unity between school mathematics and university mathematics. But with regard to the accompanying cultural environment, that is very important for the interpretation of the signs, there will always be comparably rigid differences between elementary school mathematics and university mathematics.

The notion of the cultural environment or the mathematical culture will be used as a fundamental concept for the question of what is the particular epistemological status of mathematics in the process of classroom teaching and learning. The importance of culture for both scientific and school mathematics has been emphasized a number of different occasions (Wilder, 1981, Bishop, 1988). Wilder (1986) characterizes the concept of culture as follows:

> A culture is the collection of customs, rituals, beliefs, tools, mores, etc., which we may call cultural elements, possessed by a group of people, such as a primitive tribe or the people of North America. Generally it is not a fixed thing but changing with the course of time, forming what can be called a ‚culture stream‘. It is handed down from one generation to another, constituting a seemingly living body of tradition often more dictatorial than Hitler was over Nazi Germany; in some primitive tribes virtually every act, even such ordinary ones as eating and dressing, are governed by ritual. Many anthropologists have thought of a culture as a super-organic entity, having laws of development all its own, and most anthropologists seem in practice to treat a culture as a thing in itself, without necessary referring (except for certain purposes) to the group or individuals possessing it (Wilder, 1986, p. 187).

The use of symbols, as well as the way of their reading and interpretation, is particular in every culture.

> Without a symbolic apparatus to convey our ideas to one another, and to pass on our results to future generations, there wouldn‘t be any such thing as mathematics – indeed,

> there would be essentially no culture at all, since, with the possible exception of a few simple tools, culture is based on the use of symbols. A good case can be made for the thesis that man is to be distinguished from other animals by the way in which he uses symbols.... (Wilder, 1986, p. 193).

Also for understanding the interactive development of mathematical knowledge in the classroom, it is both appropriate and productive to interpret the reciprocal relations between mathematics and teaching mathematics, and the interaction between the teacher and students, as events of a particular culture.

> Symbols, symbolic relationships and the introduction into the use and the reading of symbols are essential aspects for the formation of every culture (cf. Wagner 1981; 1986). Mathematics deals *per se* with signs, symbols, symbolic connections, abstract diagrams and relations. The use of the symbols in the culture of mathematics teaching is constituted in a specific way, giving social and communicative meaning to letters, signs and diagrams during the course of ritualized procedures of negotiation. Social interaction constitutes a specific teaching culture based on school-mathematical symbols that are interpreted according to particular conventions and methodical rules. In this way, a very school-dependent, empirical epistemology of mathematical knowledge is created interactively, in which the reading of the symbols introduced is determined strongly by conventional rules designed to facilitate understanding, but, in this way, entering into conflict with the theoretical, epistemological structure of mathematical knowledge. The ritualized perception of mathematical symbols seems to be insufficient according to the analysis of critical observers, because it does not advance to the genuine essence of the symbol but remains on the level of pseudo-recognition (Steinbring, 1997, p. 50).

Wilder makes use of observable differences in the application of symbols for explaining „good“ and „bad“ mathematics teaching.

> Man possesses what we might call symbolic initiative; that is, he assigns symbols to stand for objects or ideas, sets up relationships between them, and operates with them as though they were physical objects. So far as we can tell, no other animal has this faculty, although many animals do exhibit what we might call symbolic reflex behavior. Thus, a dog can be taught to lie down at the command ‚Lie down,‘ and of course to Pavlov‘s dogs, the bells signified food.... However, much of our mathematical behavior that was originally of the symbolic initiative type drops to the symbolic reflex level. This is apparently a kind of labor-saving device set up by our neural systems. It is largely due to this, I believe, that a considerable amount of what passes for ‚good‘ teaching in mathematics is of the symbolic reflex type, involving no use of symbolic initiative. I refer of course to the drill type of teaching which may enable stupid John to get a required credit in mathematics but bores the creative minded William to the extent that he comes to loathe the subject! (Wilder, 1986, p. 193/4).

Over the last few years, classroom research on the learning of mathematics especially has taken an ethnographic perspective towards mathematical classroom interaction and, as such, interpreted mathematics teaching as a special *culture* (Cobb & Yackel, 1998; Nickson, 1994). An essential point of this approach is not to define, by means of a scheme given a priori, what is good mathematics instruction and how good teaching material must be designed. And then, according to this approach, differences and disparities among assumptions, aims, and the classroom practice observed are examined. On the contrary, mathematics instruction is conceived as a distinct culture in which the understanding and development of knowledge take place in an autonomous and self-referential way (cf. Bauersfeld, 1982; 1998; Voigt, 1998).

> Participating in the process of a mathematics classroom is participating in a culture of mathematizing. The many skills, which an observer can identify and will take as the main performance of the culture, form the procedural surface only. These are the bricks of the building, but the design of the house of mathematizing is processed on another level. As it is with culture, the core of what is learned through participation is *when* to do *what* and *how* to do it. ... The core part of school mathematics enculturation comes into effect on the meta-level and is ‚learned' indirectly (Bauersfeld, cited according to Cobb, 1994, p. 14).

Mathematical knowledge cannot be reduced to signs, symbols, principles, and types of structures. The interpretation of the mathematical symbols requires a cultural environment. Furthermore, as will be pointed out, the understanding of mathematical knowledge and mathematical concepts cannot be sufficiently explained only within the frame of cultural behavior, usage and interpretation. Mathematical knowledge is *theoretical* knowledge. This means that mathematical concepts cannot be logically derived from, nor completely traced back to, other mathematical concepts (This idea will be explained in more deepth with the help of the probability concept later in this section). This argument is based on the understanding that a mathematical concept is not to be identified with its coherent, formal definition but encapsulates a multitude of emerging meanings. Mathematical concepts represent a relatively autonomous epistemological entity. In order to have students understand mathematical concepts, a cultural environment is necessary but it is not sufficient alone. Furthermore, the learners must behave actively in this cultural environment and must detect possible interpretations of the mathematical concept. Mathematics has to be understood as an activity (Freudenthal). When performing a mathematical activity, the learners have to be aware of the epistemological particularity of the mathematical knowledge in their process of interpreting and understanding and, in this way, they themselves have to construct the new mathematical relations.

A central criterion of theoretical mathematical knowledge – also observable in the course of its historical development – lies in the transition from pure object, or substance thinking, to relation or function thinking.

> Every one knows that there are things, and relations between things. Classical mathematics is primarily and essentially a *thing*-mathematics. The Western is primarily and essentially a *relation*-mathematics, a mathematics of functions – that is, relations –, in accord with the Faustian saying of Henri Poincaré, that science can not know ‚things' but only ‚relations' (Keyser, 1932/33, p. 193).

The transition from a *substance concept* to a *relational concept* is a central part of Ernst Cassirer's epistemological philosophy.

> ... the theoretical concept in the strict sense of the word does not content itself with surveying the world of objects and simply reflecting its order. Here the comprehension, the ‚synopsis' of the manifold is not simply imposed upon thought by objects, but must be created by independent activities of thought, in accordance with its own norms and criteria (Cassirer, 1957, p. 284).

And in another passage, Cassirer writes:

> It is evident anew that the characteristic feature of the concept is not the ‚universality' of a presentation, but the universal validity of a principle of serial order. We do not isolate any abstract part whatever from the manifold before us, but we create for its members a definite relation by thinking of them as bound together by an inclusive law (Cassirer, 1923, p. 20).

It is the theoretical relation which is characteristic of the concept. The *relation* for the constitution of relational or functional concepts takes the place of the *things* on which the substance concepts are founded. To give a clarifying example: In the „world of objects", „0" (zero or null) means „no object": and in this world there is no principal difference between the removal of „5 apples and 5 pears" or of „5 black and 5 red chips". If, in the model with black and red chips, the same number of black and red chips is given to *mean* „0", this theoretical relation has to be established „by ones own and independent activities of thinking" and, only in this way, a difference is constructed between the chips configuration, which *symbolizes* a number aspect, and the pears and apples, which belong to the world of things.

This understanding of theoretical mathematical concepts as referring to relations, and not to objects or to the empirical properties of objects, constitutes the basic step towards developing mathematics education into a scientific discipline.

> For didactics, for instance, it is obvious that the didactic problem in its deeper sense, that is in the sense that it is necessary to work on it scientifically, is constituted by the very fact that concepts will reflect relationships, and not things. Analogously, we may state for the problem of the application of science that it will become a real problem only where the relationship between concept and application is no longer quasi self-evident, but where to establish such a relationship requires independent effort (Jahnke & Otte, 1981, pp. 77/78).

Mathematical concepts are not empirical things, but represent relations. Raymond Duval explains this position as „the paradoxical character of mathematical knowledge":

> ... there is an important gap between mathematical knowledge and knowledge in other sciences such as astronomy, physics, biology, or botany. We do not have any perceptive or instrumental access to mathematical objects, even the most elementary, ... We cannot see them, study them through a microscope or take a picture of them. The only way of gaining access to them is using signs, words or symbols, expressions or drawings. But, at the same time, mathematical objects must not be confused with the used semiotic representations. This conflicting requirement makes the specific core of mathematical knowledge. And it begins early with numbers which do not have to be identified with digits and the used numeral systems (binary, decimal) (Duval, 2000, p. 61).

On the one hand, mathematical objects require signs and symbols in order to code the knowledge and to operate with this knowledge. However, attention must be paid since these signs are not identical with the mathematical objects or relations. How are the signs and symbols, used in the context of a mathematical classroom culture, related to „objects"? Bauersfeld criticizes the assumption that „names / words correspond to the matching objects", an assumption long self-evident in the philosophy of language (Bauersfeld, 1995, p. 273). Henceforth it is questionable that the objects / things, which the signs and symbols refer to, are independently preexisting, as well as whether the symbols are mere names for objects. There is no easy correspondence between objects and symbols. The dominance of a supposed easy correspondence between objects and symbols is one reason for the failure of traditional teaching methods.

> The teacher knows and teaches the truth, using language as a representing object and means. Because there is no simple transmission of meaning through language, the students all too often learn to say by routine what they are expected to say in certain defined situations (Bauersfeld, 1995, p. 275/6).

The particular interrelationship between „signs / symbols" and „objects / references" is central for the description and analysis of mathematical classroom interaction as a specific culture. Furthermore, this relation represents the core of the epistemologically-oriented analysis of mathematical interactions used in the following discussion (see chapter 3). Every mathematical knowledge requires *certain sign or symbol systems* to grasp and code the knowledge. In elementary teaching, these are mainly arithmetical number signs. These signs, on their own, do not have a meaning. The meaning of a mathematical sign has to be constructed by the student. The form and structure of the signs have emerged historically and are largely conventional and arbitrary. Generally speaking, in order to obtain meaning, mathematical sign systems require suitable *reference contexts*. Meanings for mathematical concepts are actively constructed by the learner or the teacher as interrelationships between sign / symbol systems and reference contexts / object fields (Steinbring, 1993).

As a first approach, the characterization of the role of *mathematical signs* requires consideration of two functions:

(1) A semiotic function: the role of the mathematical sign as „something which stands for something else".

(2) An epistemological function: the role of the mathematical sign in the context of the epistemological interpretation of mathematical knowledge.

According to the semiotic function, the mathematical sign stands in relationship to „something else", thus to an opposing object, usually called the reference object (cf. Nöth 2000, p. 141). The suggested relation of object and sign

object / reference context ⟷ sign / symbol

raises different questions: Which role do mathematical signs have, with regard to the objects to which they refer? What kind of mediation is seen between the mathematical signs and the objects? In the case of the epistemological function, the influence of the epistemological characteristics of mathematical knowledge comes into play. „Epistemology is about the relationship between these types of entities, objects and signs" (Otte, 2001, p. 3). Depending on the conception of the nature of mathematical knowledge, different interpretations are connected with the mathematical signs. An empirical foundation of mathematical knowledge will appoint a status to mathematical signs different from, for example, a theoretical conception of mathematical knowledge. In the following, the specific kind of the mediation between „object / reference context" and „mathematical sign / symbol" shall be elaborated. According to the theoretical position taken here, this specific mediation is a core element of the epistemology of mathematical knowledge.

The distinction made in the characterization „signs / symbols" is a consequence of the following consideration. Primarily, „sign" is to be understood as a given material object (of different forms, e.g. as a cipher, a letter, an icon, a diagram, or a painting, a gesture, a concrete thing, etc.), where this sign is important, not as an object, but with regard to its function, that it stands for or is in reference to, something else. In this meaning, objects used daily, such as traffic signs, are signs that refer to something else, for example, to a stopping restriction, a traffic jam, or to a stop. The difference between sign and symbol shall be defined as follows: First, symbols also have the function of a sign, i.e. they refer to something else. For example, the symbol 17.05 € represents a certain amount of money or a price with respect to the value of an object. A (mathematical) symbol is mainly characterized by the property that it possesses in itself an internal relational structure. The symbol 17.05 € has such a structure, given to it by the decimal positional system, but the traffic sign STOP does not indicate such a structure. Consequently, the double characterization as „sign / symbol" shall express the possible difference in the use of mathematical „notations": in a first and direct use as signs referring to something else and in an extended, deeper perception as a symbol having its own relational structure.

For working out the particularities of mathematical signs in connection to the two dimensions mentioned above, the so-called epistemological triangle can be used as a conceptual scheme (cf. Steinbring, 1989; 1997). Mathematics requires *certain sign* or *symbol systems* to record and codify knowledge. The outer form of mathematical signs has developed historically and is largely laid down conventionally (cf. for the number signs for example Menninger, 1979). To start with, these signs do not immediately have a meaning of their own. The meaning has to be produced by the student or the teacher by establishing a mediation between *signs / symbols* and suitable *reference contexts*. This mediation is not entirely subjective and arbitrary, but, in order for the signs to become true mathematical signs, their connections to possible reference objects is determined by epistemological conditions of mathematical knowledge as is evident for elementary mathematical concepts (Steinbring, 1991b). The triangular connecting scheme between the mathematical signs, the reference contexts, and the mediation between signs and reference contexts, which is influenced by the epistemological conditions of mathematical knowledge, can be represented in the *epistemological triangle* (cf. Steinbring, 1989; 1991a; 1998).

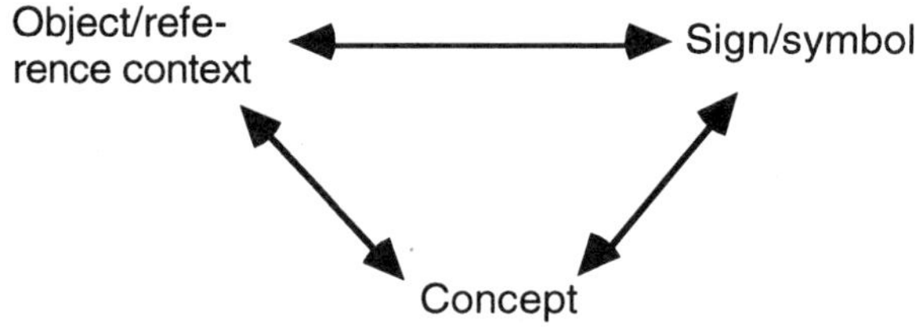

Figure 2. The epistemological triangle.

By means of the epistemological triangle, a semiotic mediation between „sign / symbol" and „object / reference context" is modeled, which is simultaneously formed by the epistemological conditions of mathematical knowledge. Furthermore, one has to note that the epistemological constraints of the mathematical knowledge not only exert influence on the mediation between sign and reference context, but that, at the same time, new and more general mathematical knowledge can be constructed through mediations between signs and reference contexts. Accordingly, none of the points of this triangle is explicitly given or definitely fixed a priori in such a way that one of the three points could become a secure starting place for definitely determining the triangle. The three reference points „mathematical concept", „mathematical sign / symbol" and „object / reference context" constitute a balanced and reciprocally supportive system.

Furthermore, this triangular scheme is not seen as independent from the student or the teacher. The reciprocal actions between the „points" of the triangle and the necessary structures for the signs / symbols (for example mathematical operations) and the object / reference context (for example diagrams, functional structures etc.) must be actively produced by the student (in the interaction with others and with the teacher). This active production is always subject to the epistemological constraints. Thus the epistemological triangle serves to model the nature of the (invisible) mathematical knowledge by means of representing the relations and structures constructed by the learner in the interaction. Furthermore, one can accordingly draw up a sequence of epistemological triangles for the interaction, or a sequence of learning steps to reflect the development of interpretations made by the subject. The student and their activity in social interaction is not an analytically equal element at the level of the three components of the epistemological triangle, but is a kind of meta-element responsible for the construction of relations in this scheme. In the ongoing development of mathematical knowledge, the interpretations of the sign systems and the appropriately chosen reference contexts are modified or if necessary further generalized by the student or the teacher.

For the analysis of the semiotic problem, of how relations between symbols and referents are realized, similar triangular schemes have been developed in mathematics, linguistics and the philosophy of language. A comparable triangle has been introduced by Frege (1892, 1969): „sign, sense, and meaning". In this scheme „meaning" represents the objective idea of the thing. „Sense" contains the subjective interpretation which a person undertakes with the object and which is related to the „meaning", but is subject to changes. And „sign" is a name for designating the objective idea. One of Frege's central assumptions states that the objective idea, the „meaning", exists independently beforehand, and that, accordingly, the other elements of the triangle are determined by that „meaning".

A further triangle of meaning (of signs and symbols) was developed by Ogden and Richards (1923; cf. also Steinbring, 1998a). They explain the problem of the „meaning of the meaning" as follows.

> Many difficulties indeed, arising through the behavior of words in discussion, even among scientists, force us at an early stage to take into account these ‚non-symbolic' influences. But for the analysis of the senses of ‚meaning' with which we are chiefly concerned, it is desirable to begin with the relations of *thoughts*, *words* and *things* as they are found in cases of reflective speech uncomplicated by emotional, diplomatic, or other disturbances; and with regard to these, the indirectness of the relations between words and things is the feature which first deserves attention (Ogden & Richards, 1923, p. 10).

Ogden and Richards describe this structure of „thoughts," „words," and „things" in a triangular diagram (Fig. 3).

> Between the symbol and the referent there is no relevant relation other than the indirect one, which consists in its being used by someone to stand for a referent. Symbol and Referent, that is to say, are not connected directly (and when, for grammatical reasons, we imply such a relation it will merely be an imputed, as opposed to a real relation) but indirectly round the two sides of the triangle" (Ogden & Richards 1923, p. 11–12). And later: „The root of the trouble will be traced to the superstition that words are in some way parts of things or always imply things corresponding to them, ... The fundamental and most prolific fallacy is, in other words, that the base of the triangle given above is filled in (Ogden & Richards, 1923, p. 14–15).

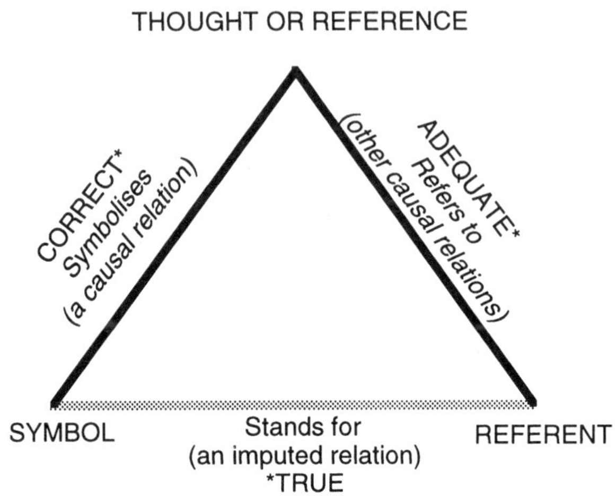

Figure 3. The triangle of meaning.

Ogden and Richards emphasize that the relation between symbol and referent is not given in a pre-fixed manner, but is of an indirect nature, and thus this relation has to be constructed in an agreed way. They conceive of this construction, in principle, as a relation between „symbol“ and „referent“ by means of the „thought or reference“, as a kind of definition process in which one can achieve a relatively precise definition of empirical objects by social conventions and logical correspondences.

In the epistemological triangle, other than in Frege‘s scheme, the reference point of „object / reference context“ is not to be understood as an independent, pregiven element. And, in contrast to the triangle of meaning by Ogden and Richards, the construction of relations between „sign / symbol“ and „object / reference context“ over the „concept“ does not lead to final, unequivocal definitions, but is understood as a complex interrelationship. As explained before, the connections between the corners of the triangle are not explicitly defined and unchangeable. They form a mutually supporting system in which the interpretations of „object / reference context“, „sign / symbol“ and „concept“ do not occur locally, but globally, by reciprocal actions within the system. In the course of further development of mathematical knowledge, the interpretation of the sign system with the matching reference contexts will change.

The development of the probability concept can be used as a pradigmatic example for explaining essential features of the epistemological triangle. In the early history of the probability concept the sign system is given by „fraction numerals“ and an accompanying reference context is given by the „ideal die“. Later in history, the reference context changed to „statistical collectives“ (cf. Mises, 1972), and the sign system described the „limit of relative frequency“ by a mathematical expression (as for instance $h_n(E) \approx p(E)$ for a rather big number n of trials). And, at the beginning of the 20th century, the reference context changed to „stochastically independent / dependent structures“ and the sign system listed mathematical statements representing „implicitly defined axioms“ (cf. Steinbring, 1980).

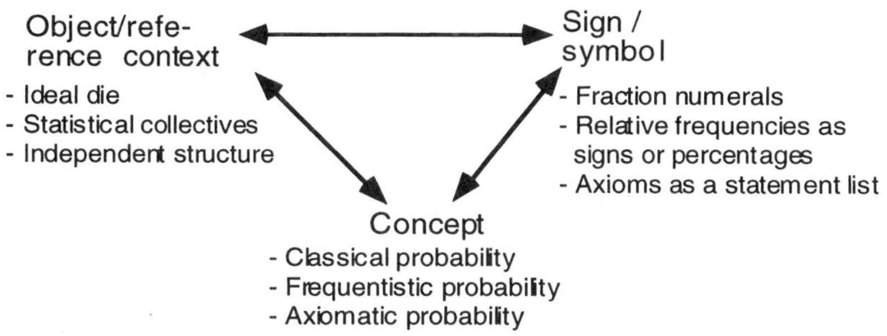

Figure 4. The epistemological triangle applied to the concept of probability.

A central characteristic of this semiotic structure is the fact that the object or the reference context cannot be a fixed and definite point, but that it is interpreted by the learner more and more as a structural domain during the development of mathematical knowledge. Accordingly, mathematical meaning is produced in the interplay between a reference context and a sign system, by means of transferring possible meanings from a relatively familiar, or partly known, reference context to a new, still meaningless, sign system.

This fundamental way of creating mathematical meaning for the number signs takes place from the beginning of mathematics learning in the elementary classroom. In early mathematical teaching, so called real world problems and real world pictures are often chosen as reference contexts.

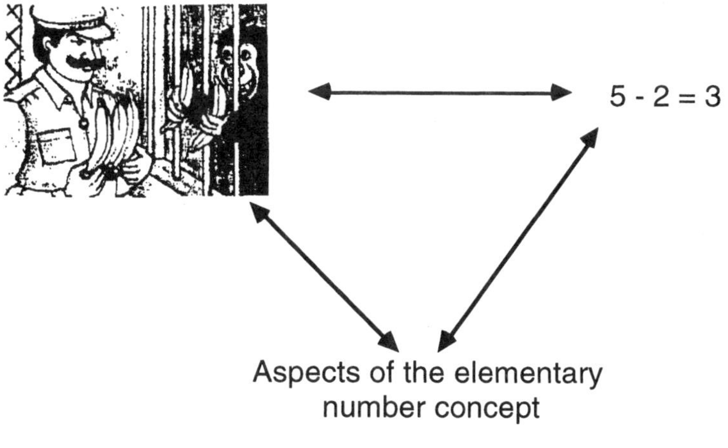

Figure 5. The epistemological triangle with an empirical reference context.

This real world picture, for example, of „guard and monkey“ is typical of the design of reference contexts widely found in elementary textbooks. These pictures are standardized, and the children quickly learn to correctly interpret them mathematically in the expected way. (cf. Neth & Voigt, 1991; Voigt, 1991). Many children decode such „didactified" real world pictures according to characteristic attributes: burnt matches pointing to the right symbolize a minus-exercise, as do eaten apples, cracked nuts or birds flying away. But joining children, or balls running close, point to a plus-exercise. Multiplication exercises can be quickly recognized with their rectangular shelves, bottle-cases or egg-boxes, and, when finally a number of apples is to be packed into paper bags, it can only be about dividing or distributing.

Reference contexts arranged this way become standardized picture patterns, which are mostly not concerned with exploring meaningful circumstances. These standard pictures primarily contain isolated clues of recognition and support an empirical interpretation of the arithmetical operation: *addition* is coming *here*, jumping *here*, putting *here* etc., *subtraction* is taking *away*, going *away*, flying *away*, eating, burning, etc. And, accordingly, empirically definite descriptions exist for *multiplication* and *division*. In this way, a kind of directness between the sign systems and the reference contexts is intended, which is supposed to make learning easier and make a direct creation of meaning possible (Voigt, 1991).

Such a didactified form reduces the relation between the „sign system" and the „reference context" to a limited single reading. For example, this reduction does not offer sufficient possible structures where the children are expected to understand different arithmetic relations and strategies meaningfully and use them in a flexible manner. In the course of unfolding the number concept, the relation between the „signs / symbols" and the „reference contexts" changes. The empirical character of knowledge is increasingly replaced for the benefit of a relational connection between the number-signs and the reference contexts. Concrete and empirical things are superseded by diagrams and means of illustration, or by structure sign systems of a new kind.

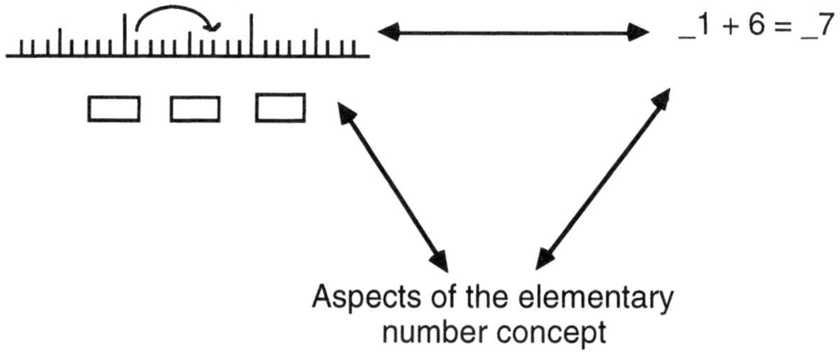

Figure 6. The epistemological triangle with a relational reference context.

In the above example (Fig. 6) a first interpretation could be, that the diagram showing a part of the number line, represents the „reference context" and the arithmetical task „_1 + 6 = _7" might express the yet unknown system of „signs / symbols". The relational structure that is embodied in both domains is important for the mediation between the „signs / symbols" and the „reference context". In addition, some students also might interpret the roles of the number line and the arithmetical task in another way. The number line could be an unknown symbolic diagram, whereas the arithmetical task might be known to a certain extent and could be used as a rather familiar „reference context" for explaining the unknown diagram. This different interpretation of the epistemological triangle makes clear that the reference context itself is to be interpreted as a structured system, and, in this way, interrelationships with „equal rights" between the sign systems and the reference contexts emerge.

Developing mathematical meanings in the interplay between a reference context and a sign system thus means the ability to transfer possible relations from a relatively familiar, or, in some aspects, known reference context, to a new, still meaningless, sign system. In this way, a flexible switching back and forth between reference context and sign system becomes possible, while the „roles" of the sign system and reference context become interchangeable.

Fundamental questions of mathematics education are connected with the epistemological triangle. Among other things, these questions are about acquiring a suitable concept of the epistemological nature of mathematical knowledge in processes of teaching and learning. An appropriate epistemology of mathematical knowledge can serve certain educational purposes such as the classification and analysis of school-mathematical knowledge in textbooks or in the curriculum, as well as the qualitative analysis of mathematical communication between students and teacher. In this regard, the epistemological triangle is not simply an offshoot of other similar conceptual diagrams (e.g. Frege (1892; 1969), Ogden & Richards (1923)). It is a specific theoretical instrument for mathematics education, that is used for characterizing the particularities of mathematical knowledge and for analyzing mathematical interaction and communication.

Educational problems in mathematics, as they become visible from an epistemological perspective in everyday teaching practice, have been an essential source of the development of the epistemological triangle.

> The practice of mathematics, especially in school, is ... usually reduced to an identification of sign and signified by means of the atomization, the algorithmic, as it is expressed in formulas as a procedure of calculating, or, if one makes the threefold distinction of concept, sign and object, which would in fact be necessary, it leads to an identification of sign and object, neglecting an independent concept. (Otte, 1984, p. 19).

The true mathematical object, that is, the mathematical concept, must not be identified with its representations.

> A mathematical object, such as a function, does not exist independently of the totality of its possible representations, but it is not to be confused with any particular representation, either. It is a general that ... cannot as such be exhausted by any number of representations (Otte, 2001, p. 33).

This criticism on the identification of sign and object or even with the mathematical concept, is also formulated in the philosophy of mathematics:

> ... in certain branches of mathematics the symbols and diagrams have often been confused with the very mathematical objects they are supposed to denote or represent. Thus we not only take the result of manipulating numerals or geometric diagrams or paper Turing machines as direct evidence of properties of numbers, geometric figures or abstract Turing machines, we also tend to confuse numerals with numbers, drawings with figures, and paper machines with abstract ones. Here structuralism has an advantage, because the symbol systems themselves form structures isomorphic to the structures they are supposed to represent. In so far as we ignore the identifying features (their

> shape, for example) of these symbols and focus only upon how they are related to each other, there is, for the structuralist, no fact of the matter as to whether we are contemplating notational objects or the positions in a structure (Resnik, 2000, p. 361).

The triple distinction „concept“, „sign / symbol“ and „object / reference context“ describes the fundamental structure of the epistemological triangle. The following two issues are essential for the epistemological triangle from the point of view of both mathematical knowledge and mathematic-didactical questions:

- There exists a third element independent from object and sign, the concept.

The need for an independently existing mathematical concept shall be discussed in the following with the help of the probability concept. The concept of probability has to be distinguished from the signs; but it needs the signs. For example, the probability of throwing a 5 with an ideal die can by written with the following signs: $P(\{5\}) = {}^1/_6$. And such a notation could be simply seen as „only a fraction“. This symbolic notation of ${}^1/_6$ cannot be identified with the elementary probability concept. The signs / symbols in a mathematical theory are different from the objects. In elementary probability, for example, a distinction has to be made between signs and dice or wheels of fortune, or further mainly mental objects, or random structures, and later also structured reference contexts. But the new, unfamiliar signs need some referential objects so that they can be interpreted and gain meaning. In addition, the mathematical concept is also independent from the object / reference context, but needs this „reference context“ as a „situated embodiment“ of a structure or relation. Furthermore, the concept has to be distinguished from the mediating relationship between „object – sign“. This mediation alone insufficiently expresses the conceptual aspects involved. In order to organize and evaluate the mediation between „signs / symbols“ and „reference contexts“ in probability theory , a core idea of the concept of probability must have been already used from the beginning. This dialectic between a pre existing conceptual idea of probability and the epistemological requirements for the adequate mediation between „signs / symbols“ and „reference contexts“ shall be elaborated in the following paragraphs.

This fundamental point of view, that mathematical concepts, as theoretical concepts, cannot be entirely reduced to other concepts nor to other knowledge, but eventually exist independently, became clearly and explicitly visible for probability for the first time in Bernoulli‘s theorem. In the frame of early classic probability, the signs used to code probabilities were fraction numbers which indicated the proportion of favorable to unfavorable cases. However, these „ideal“ measured values must be carefully distinguished from the „true“ probabilities of a chance experiment with random generators. With chance experiments, the probability can be estimated in a preliminary way with the help of the empirical law of large numbers, thus of observed relative frequencies.

According to the given situation in the epistemological triangle (cf. Fig. 4), one can view the ideal fraction numbers as examples of the point „mathematical signs / symbols“, in order to determine the searched probabilities. And the patterns of relative frequencies (as measured empirical values), which can be observed in the real chance experiment, can be placed under the corner point „object / reference context“.

How is a relation established between „object / reference context“ and „mathematical signs / symbols“? How is in the example of probability a relation established between empirical and ideal probability? It would be too simple to claim that the (ideal) probability eventually becomes identical to the relative frequency (after very many trials), thus having an identity formed between sign and object (Otte, 1984). However, this relation is not a simple identity, it is based on a complex structure which is essentially determined by the epistemological conditions of the probability concept. The preliminary, rather direct, relation between relative frequency and classic probability in the frame of the empirical law of large numbers, is an assertion which must itself be mathematically analyzed and described according to mathematical models and rules. In the early history of the theory of probability, the famous theorem of Bernoulli is the first exact formulation of this relation (cf. Loève, 1978).

Bernoulli's Theorem:

Let h_n be the relative frequency of 0 in n independent trials with two outcomes 0 and 1, which have the probability $p \in [0,1]$, then follows:

$$\forall\ \varepsilon > 0, \exists\ \eta > 0, \exists\ n_0, \forall\ n \geq n_0 : P(|h_n - p| < \varepsilon) \geq 1 - \eta.$$

> Altogether we have three variable quantities: first the precision of the statement considered, which is measured by e, then the certainty with which the statement holds, measured by h, and, finally, the number of trials made, which is given by n. These three parameters are mutually dependent on one another, it is possible to fix two and then to attempt to estimate the third (Steinbring, 1980, p. 131).

This statement in Bernoulli's theorem, that there is a very great probability that the relative frequency and probability of the stochastic experiment will be as close to each other as desired if the number of trials increases, leads to the following consequence: The relation between the signs (the ideal fractions for elementary probabilities) and the reference objects (the chance experiments with the observed relative frequencies) cannot be understood as a fixed and safe definition of the concept, but that the mediation between ideal and empirical probabilities can only take place on the basis of a pre-required and thus relatively independent conceptual idea of probability.

In the attempt to mediate between the ideal signs for the elementary probability and the empirical observations, i.e. the relative frequencies in the chance experiment, one has to presuppose, to a certain extent, the concept of probability which must be defined and clarified. This epistemological problem is known as the *circularity of mathematical concept definitions* (in particular for the elementary concepts of probability, cf. Borel, 1965). This circularity in the construction of mathematical concepts should not be seen simply as a defect. On the contrary, the impossibility of deducing a mathematical concept in all its details from other knowledge elements means that the necessary presupposition of autonomous central conceptual ideas, demonstrates, in the end, the autonomy of the elementary probability concept. Accordingly, the case of the elementary probability concept should serve to justify the autonomy of the point „mathematical concept“ in the epistemological triangle with regard to the other points of „object / reference context“ and of „mathematical signs / symbols“ (especially cf. Steinbring, 1980; 1991c).

• If, as happens, concrete, external objects are taken in the beginning – e.g. in elementary school – as the referring objects for signs, these will be more and more superseded by mental objects and by structures etc. in the further development of mathematical knowledge.

Many elementary school pupils soon realize that the concrete objects presented to them – such as chips, one-cubes or ten-bars – are not interesting as external, fixed objects, but because they embody mathematical ideas or structures. At this point, the student plays an important role without being explicitly mentioned in the epistemological triangle. It is the human being (the student or the teacher) that creates connections between signs and objects and, moreover, they decides how signs and objects are to be read and interpreted. At the beginning, one may start with „external, empirical objects“, but in principle, it is always about mental ideas on relations and structures – embodied in signs and reference contexts.

The sign-systems in mathematics are not mere „names or abbreviations of any objects (even abstract ones)“, they contain structures, patterns and relations. The quantity „12.371 km“, for example, is a mathematical sign with a structure which can be represented by the position table (Steinbring, 1998b). Since the sign-systems, as well as the reference contexts, essentially represent structures and relations, it becomes possible – and this also happens in the interactive processes of mathematical knowledge development – to have a „change“ of position between sign and reference contexts by subjective interpretation.

In particular, the „interchangeability of the positions of sign / symbol and reference context,“ represents an important characteristic of the epistemological triangle. Thus, given the circumstances, neither reference contexts nor signs / symbols are external nor fixed, but are mental ideas which embody structures. Furthermore, it is important to be aware that the reference contexts (objects) do not simply precede the mathematical signs, neither temporally nor logically. On the contrary, the sign-systems can, by means of their internal structure in the (historical) development, increasingly gain independence and autonomy toward the reference contexts and thus act upon the reference contexts and project a structural interpretation, on them. Sign-systems and reference contexts, then have equal rights and are equivalent on the temporal dimension with neither preceding the other. The semiotic analysis of the historical development of the number zero is a paradigm example in this sense (cf. Rotman, 1987).

The impossibility of a direct sensing perception of mathematical knowledge (cf. Duval, 2000), as it is assumed in the epistemological triangle, and the specific function of mathematical signs of relating to structural reference contexts, where this mediation is guided by epistemological conditions, require a further ongoing interpretation of the mediation between „sign / symbol“ and „object / reference context“ in the frame of mathematical knowledge. „Mathematical signs / symbols“ as well as „objects / reference contexts“ are the embodiments of not directly visible structures. The former thus do not stand for something real and not for visible objects or features. Hence, in the epistemological triangle, „sign / symbol“ as well as „object / reference context“ stand for something else, something not directly perceivable. The epistemological triangle is a model to make the invisible mathematical knowledge accessible with regard to its structural character, to describe its particularities, and to analyze interactive processes of constructing mathematical knowledge – thus, invisible relations are embodied in exemplary contexts and activities.

The particularity of a sign as a mathematical sign can be summarized by two essential aspects. (1) In the frame of mathematical knowledge, the sign as well as the reference context – that stands for the sign – are themselves always embodiments of something else, i.e. of structures and relations. (2) the mediation between sign and reference context requires the consideration of a presupposed, relatively autonomous existence of mathematical relational concept ideas, that take part in the regulation of this mediation between sign and reference context.

The possible development just described proposes different ways of how mathematical meaning is constituted. In the first instance, mathematical knowledge obtains its meaning by concrete objects and empirically existing reference contexts. In a later developmental step, the theoretical relation structure in the reference contexts comes to the fore and also a change in the assignment of the roles of „sign / symbol“ and „object / reference“ context can occur.

This interpretation of the constitution of mathematical meaning presented here receives some positive confirmation and further explanation from another perspective. In their book „Where mathematics comes from – How the embodied mind brings mathematics into being“ (Lakoff & Núñez, 2000) G. Lakoff and R. Núñez develop a foundation of mathematics in which mathematical concepts and mathematical knowledge are traced back to human everyday experiences and actions, with the help of the so-called Mathematical Idea Analysis.

> We have found that mathematical ideas are grounded in bodily-based mechanisms and everyday experience. Many mathematical ideas are ways of mathematicizing ordinary ideas, as when the idea of subtraction mathematizes the ordinary idea of distance, or as when the idea of a derivative mathematicizes the ordinary idea of instantaneous change (Núñez, 2000, p. 9).

The mathematization of general ideas happens especially in the construction and use of metaphors.

> For the most part, human beings conceptualize abstract concepts in concrete terms, using precise inferential structure and modes of reasoning grounded in the sensory motor system. The cognitive mechanism, by which the abstract is comprehended in terms of the concrete is called conceptual metaphor. Mathematical thought also makes use of conceptual metaphor, as when we conceptualize numbers as points on a line, or space as sets of points“ (Núñez, 2000, p. 6).

In this way concepts are understood as complex metaphorical networks.

> „... concepts are systematically organized through vast networks of conceptual mappings, occurring in highly coordinated systems and combining in complex ways. ... An important kind of mapping is the ... conceptual metaphor (Núñez, 2000, p. 9).

For the foundation and development of mathematical knowledge, Lakoff and Núñez emphasize the following three metaphors as particularly relevant and specific:

> First, there are grounding metaphors – metaphors that ground our understanding of mathematical ideas in terms of everyday experience. Examples include the Classes Are Container schemes and the four grounding metaphors for arithmetic.
>
> Second, there are redefinitional metaphors – metaphors that impose a technical understanding replacing ordinary concepts. ...

> Third, there are linking metaphors – metaphors within mathematics itself that allow us to conceptualize one mathematical domain in terms of another mathematical domain (Lakoff & Núñez, 2000, p. 150).

The first-mentioned grounding metaphor can be connected with an interpretation of the reference context as a domain with concrete objects and immediate experiences from which immediate meanings can be transferred to signs in a metaphorical way, so that mathematical ideas – concept aspects – can emerge. With the following mentioned redefining metaphors, immediate meanings in the concrete reference domain are redefined, for example, by formal rules and generalizing extensions of qualities, so that an increasingly structural view of the reference domain is acquired. An example is the re-definition of an empirical comparison from „more than" and „less than", to the theoretical relation „greater than" and „smaller than".

In elementary probability, the concept of stochastic independence can be seen as such a redefining metaphor: Concrete understanding of independence as „not influenced" or „no reciprocal actions" is newly defined by a mathematical rule: Two events A and B are called stochastically independent by definition if the rule of multiplication holds true: $P(A \cap B) = P(A) \cdot P(B)$. The intuitive interpretation, that the multiplication rule follows for two independent events, is replaced by saying „Two events A and B are stochastically independent if the multiplication rule is valid".

> The linking metaphors are in many ways the most interesting of these, since they are part of the fabric of mathematics itself. They occur whenever one branch of mathematics is used to model another, as happens frequently. Moreover, linking metaphors are central to the creation not only of new mathematical concepts but often of new branches of mathematics (Lakoff & Núñez, 2000, p. 150).

Linking metaphors connect one mathematical domain to another and thus refer to the fact that the source domain (of the metaphor) – in our terminology, the reference domain – is a structured mathematical domain, which bestows meaning on another structural mathematical domain, the target domain (of the metaphor) – in our terminology, the signs / symbols. In this way, the interchangeability of the roles of reference context and signs / symbols caused by a learner or the teacher may be observed as described in the context of the epistemological triangle, if the relational and structural characteristics for mathematical knowledge are further emphasized.

An essential characteristic of mathematical knowledge, as compared to other theoretical knowledge, is certainly its unity and coherence which should, however, not be understood as objective or a priori, but which occurs through negotiation between mathematicians in the frame of the „rules" of formal argumentation and proofs. In opposition to the professional practice of the mathematical researcher, learners and teachers of mathematical knowledge are exposed to a variety of possible interpretations and understandings of mathematical knowledge, caused by cultural conditions and concrete reference contexts for mathematical knowledge. Learning mathematics requires seeing mathematics as a process, while mathematical science stresses the uniform mathematical product.

Mathematical learning processes, like any culture, are about interpreting signs in an adequate manner – and mathematics is a typical science which particularly deals with signs and symbols. The epistemological triangle supports the understanding of the particularities of the interpretation and usage of mathematical signs and symbols, which relate to reference contexts and are, at the same time, regulated in their development by aspects of independent mathematical concepts. In his knowledge development, the learner makes interpretations of the possible roles the conceptual elements of the epistemological triangle could play. On the basis of a growing accentuation of structures and relations that are embodied in reference contexts and in sign / symbol systems and that are more and more refined, in particular, the roles of signs / symbols and reference contexts can be inter-changed in the interactive learning processes in different ways, forming new sign interpretations when attempting to solve mathematical problems or when constructing new knowledge.

2. THEORETICAL FOUNDATIONS AND METHODS OF EPISTEMOLOGICALLY-ORIENTED ANALYSIS OF MATHEMATICAL INTERACTION

The empirical approach and the epistemologically-oriented qualitative analysis of mathematical classroom interactions, elaborated and used in this research, belongs to the domain of interpretative classroom research of mathematical interactions (particularly cf. Cobb & Bauersfeld, 1995).

The theoretical perspective of the qualitative analyses of the interaction of everyday mathematics instruction (in mathematics education in Germany) emerged from, among other things, the criticism of, and the turning away from the paradigm of traditional subject matter based didactics called Stoffdidaktik (Steinbring, 1998c). In Stoffdidaktik, neither the mathematical classroom interaction nor the learner play essential, theoretically reflective roles, whether from a constructive or an analytic position. Stoffdidaktik considers the learner or the everyday mathematics instruction only for the purpose of pointing at the inadequacy of real teaching or real learning processes compared to ideal teaching and ideal models of understanding, as they are established in this approach. It was only around 1975 that everyday mathematical teaching and learning started to be taken seriously as processes in their own right and analyzed from an interactionist perspective (e.g. Bauersfeld. 1978; 1988; Jungwirth. 1994; Krummheuer, 1984; 1988; Maier & Voigt, 1991; 1994; Voigt, 1984; 1994).

As a result of a criticism of Stoffdidaktik, the interactionist perspective relies mainly on two (until then neglected) basic aspects: the *learning child* (in the classroom) and the *interaction between the learner and the teacher*. In this research context, one has to distinguish between two theoretical perspectives:

> The one is an individual-psychological perspective which emphasizes the learner's autonomy and his cognitive development and which leads to the concept of student-oriented, ‚constructivistic' mathematics instruction. The other is a collectivistic perspective which criticizes the ‚child-centered ideology' of the first perspective and understands learning mathematics as the socialization of the child into a given classroom culture... (Voigt, 1994, p. 78).

These two research perspectives are thus based on reference to different scientific disciplines. The individual-psychological perspective relies, for example, on cognitive psychology as well as on radical constructivism (von Glasersfeld, 1991); and the collectivistic perspective uses sociological and ethnographic theories. In the analyses of mathematical interactions, one or the other of these two theoretical orientation is often emphasized. (Concerning the individual-psychological perspective see e.g. Cobb, Yackel & Wood (1991); and for the collectivistic perspective see Solomon (1989)). An overemphasis of either the individual-psychological or the collectivistic perspective was a major critique and a starting point for the working group around H. Bauersfeld to develop a theoretical concept explicitly bringing together the individual, cognitive perspective and the collective, social perspective as a basis for qualitative analyses of interaction.

On the one hand, it is stated that a single student cannot discover all school knowledge by himself. „Culture, we can say, is not discovered; it is traded or falls into oblivion. All this indicates for me that we should rather be more careful when talking about the discovery method or about the conception that discovery is the basic vehicle of instruction and education" (Bruner, 1972, p. 85). On the other hand, it is considered doubtful that effective participation in social interaction patterns can lead to successful mathematics learning.

> In everyday lessons interaction patterns often can be reconstructed in which the teachers influence every step of the students' activities without creating favorable conditions for the student to make desirable learning processes in problem solving and developing concepts.... We should resist the temptation of identifying learning mathematics with the student's successful participation in interaction patterns (Voigt, 1994, p. 82).

Consequently, an interaction theory has been developed, in which both perspectives were connected to each other.

> [A]n interaction theory of teaching and learning mathematics [offers] a possibility of regarding social aspects of learning mathematics and at the same time of avoiding the danger of overdoing the cultural and social dimensions. For the interaction theory emphasizes the processes of sense making of individuals which interactively constitute mathematical meanings. The interaction theory of teaching and learning mathematics uses findings and methods of microsociology, particularly of symbolic interactionism and ethnomethodology (cf. Bauersfeld, Krummheuer & Voigt, 1988; Krummheuer & Voigt, 1991). Of course the interaction-theoretical point of view does not suffice if one wants to understand classroom processes holistically (Voigt, 1994, p. 83).

Regarding the interaction-theoretical perspective, the research approach of the *social epistemology of mathematical knowledge* used in this work understands itself as an important, independent complet model inasmuch as the particularity of the social existence of mathematical knowledge is an essential component of this theoretical approach of interaction analysis. The conceptual emphasis in the theoretical perspectives portrayed so far was focused on the psychological or social processes involved in mathematical interactions and the specifics of mathematical knowledge remained unconsidered due to ignoring Stoffdidaktik. In the theoretical conception of the social epistemology of mathematical knowledge, the epistemological particularity of the subject matter „mathematical knowledge" dealt with in the interaction constitutes a basis for its theoretical examination. In this theoretical investigation mathematical knowledge is seen from a different perspective: the subject matter of „mathematics" is, according to the considerations in the previous section, not understood as a pregiven, finished product, but interpreted according to the epistemological conditions of its dynamic, interactive development.

Every qualitative analysis of mathematical communication always has to start – explicitly or implicitly – from assumptions about the status of mathematical knowledge. There are different ways of coping with this requirement. There could be a general assumption, to observe and to analyze mathematics teaching in the same way as every other form of teaching, without taking into account the particularities of mathematical knowledge. Such a position negates a possible influence of the content of the instruction on the qualitative analysis. Or another, typical, assumption could be that mathematical communication is, indeed, determined by the specific subject matter dealing with an „objective", correct subject matter knowledge, and therefore the analysis of mathematical communication could reach an unequivocal assessment as „true / false" or „good / bad teaching"

A fundamental assumption for different research approaches to mathematical interaction is the idea that the mathematical subject matter cannot be introduced into the teaching / learning process as a ready made curricula product, but that the subject matter knowledge can only be mutually generated during the interactive process. This assumption contains the following contradiction: Teaching is an activity oriented towards a pre given goal which requires the step by step administration of subject matter to students. In contrast to this, learning is seen as an active process of construction and development, which, through interaction, is the basis for the emergence of new knowledge. When starting from this supposition the central objective for interpretative research is to reconstruct and understand interactive knowledge development in the mathematics classroom as an evolving autonomous process dependent on internal conditions. Ready made mathematical knowledge as seen by the teachers can no longer be the seemingly objective measure according to which the success or failure of the teaching-learning process could be read. On the contrary, the intention of achieving an instructional goal becomes, in itself, a factor which influences the interactive construction of knowledge and can thus stimulate or hinder it.

In the course of interaction analysis, interpretations must be made of the verbal and non-verbal communications of those interacting. These qualitative interpretations have to be consistent in some way; often there will not only be one „true" explanation but several, alternative, plausible interpretations of analyzed communications (cf. Voigt, 1994). These interpretative analyses bring to light „typical" patterns of interaction; and comparative analysis of several „similar" teaching episodes can provide a more secure data base for decisions about the chosen interpretation of observed communication (Krummheuer, 1997).

When educational researchers interpret the statements observed in classroom communication, this interpretation is influenced by assumptions about the role of mathematical knowledge. For instance, there is the belief that all interactions that are related to mathematics create mathematical knowledge per se and that therefore the interactively emerging mathematical knowledge can only be reconstructed in the course of the analysis. This implies that there is no mathematical knowledge independent from interactions and communications. Thus the interpretative reconstruction could not and should not be undertaken with the idea of an objectively existing, correct mathematical knowledge in the background. Such an idea would not do justice to the particularities of the social situation, and the interaction analysis would be pressed into an external, artificial evaluation framework. The interactive mathematical situation is supposed to be conceived and reconstructed only out of itself.

Epistemology-based interaction research in mathematics education proceeds on the assumption that a specific social epistemology of mathematical knowledge is constituted in classroom interaction and this assumption influences the possibilities and the manner of how to analyze and interpret mathematical communication. This assumption includes the view of mathematics explained above: Mathematical knowledge is not conceived as a ready made product, characterized by correct notations, clear cut definitions and proven theorems. If mathematical knowledge in learning processes could be reduced to this description, the interpretation of mathematical communication would become a direct and simple concern. When observing and analyzing mathematical interaction one would only have to diagnose whether a participant in the discussion has used the „correct" mathematical word, whether he or she has applied a learned rule in the appropriate way, and then has gained the correct result of calculation, etc.

The epistemology-based interaction research approach understands mathematical knowledge and mathematical concepts neither as concrete, material objects, given *a priori* in the „external" reality, nor as independently existing (platonic) ideas. For the individual cognitive agent mathematical concepts are „mental objects" (Changeux & Connes, 1995; Dehaene, 1997) and in the course of communication they are constituted as „social facts" (Searle, 1995) or as „cultural objects" (Hersh, 1997). From an evolutionary point of view, mathematical concepts develop as cognitive and social theoretical knowledge objects in a contrary relation to the material and social environment.

Other than with objects constructed by human beings such as a chair, table, knife or screwdriver, the meaning of social facts such as money, time (measured by a clock) or the number concept is not deducible neither from the form nor from the material of these objects. For example, one cannot derive the specific item of the accompanying mathematical object either from the „material" or from the functional form of the number signs –3, 17, $\sqrt{2}$, or π. The mathematical objects remain invisible in a certain sense. Their meaning is constructed by the subject – individually or in social interaction – in confrontation with experience-bound and abstract reference contexts. Thus in the epistemologically-oriented research on mathematical interaction mathematical concepts are understood neither as objective, pre-given entities, nor as exclusively subjectively produced conceptions. Mathematical concepts and mathematical knowledge are seen as socio-historically constructed „social facts" which are conceived of as „symbolized, operative relations" between their abstract codings and the socially intended interpretations.

Mathematical concepts are constructed as symbolic relational structures and are coded by means of *signs and symbols*, that can be combined logically in mathematical operations. This interpretation of mathematical knowledge as „symbolic relational structures that can be consistently combined" represents an assumption which does not require a fixed, pre-given description for the mathematical knowledge (the symbolic relations have to be actively constructed and controlled by the subject in interactions). Further, certain epistemological characteristics of this knowledge are required and explicitly used in the analysis process; i.e. mathematical knowledge is characterized in a consistent way as a structure of relations between (new) symbols and reference contexts.

The intended construction of meaning for the unfamiliar, new mathematical signs, by trying to build up reasonable relations between signs and possible contexts of reference and of interpretation, is a fundamental feature of an epistemological perspective on mathematical classroom interaction. This intended process of constructing meaning for mathematical signs is an essential element of every mathematical activity whether this construction process is performed by the mathematician in a very advanced research problem, or whether it is undertaken by a young child when trying to understand elementary arithmetical symbols with the help of the position table. The focus on this construction process allows for viewing mathematics teaching and learning at different school levels as an authentic mathematical endeavor.

The situated, interactive development of mathematical knowledge and mathematical meaning is described with the help of the epistemological triangle (section 1.1). This triangle is also used to analyze the interpretation of the roles of „sign / symbol“ and „referents“ or of the relations between them which are produced by the teacher and the students in different ways. In the course of their social discourse the students and their teacher interactively create links between the signs / symbols and appropriate reference contexts and thus they construct specific „symbolic relational structures“ in school-mathematics knowledge. On the one hand, these interactively produced structures contain constitutive epistemological conditions of the mathematical knowledge itself, as a socio-historically grown cultural knowledge domain. On the other hand, this interactive construction of knowledge structures also contains the common and individual interpretations that are negotiated in the concrete, specific teaching learning situation.

A qualitative, epistemological analysis of school-mathematical knowledge, examines and describes the tension between the socio-historically emerged structures of mathematical knowledge and the particular conditions of the specific social situation of the classroom and of learning. The scientific interest concentrates on the conditions of the social constitution of relations between sign / symbol and referents in mathematical teaching learning situations. According to the research conception presented here the epistemological triangle is used as a central instrument of description and analysis of the interactively constructed knowledge (Steinbring, 1989; 1991a; 1991b).

All processes of mathematical interaction and understanding that are used here as examples are analyzed with the help of the epistemological triangle. The focus of analysis is on the different kinds of how to understand unfamiliar „signs / symbols“ by means of producing a referential relation for an explainatory „object / reference context“. The constitution of such a relation between „object / reference context" and „sign / symbol" can be done in different ways:

- explicitly or implicitly, i.e. it can be directly presented or it can be demonstrated or it can only be „silently present“ in the background of the interactive context.
- individually or interactively, i.e. it can be created by a single person or interactively constituted.

Besides the way in which the relation „object / reference context“ <-> „sign / symbol“ is constituted, the analysis has to take into account the fact that the persons participating in the communication are usually not explicitly aware of which aspects of the knowledge play the role of the rather familiar „object / reference context“ and which play the role of the rather new and strange „sign / symbol“. Different participants can proceed on different assumptions and try to interpret unfamiliar sign systems with different familiar reference contexts. Only in the course of interactive communication growing agreement concerning the roles of „object / reference context“ and „sign / symbol“ may be observed to some extent between the participants. But due to the development of new knowledge and the negotiation of new meanings these agreements are often limited and they are questioned and changed over and over again. (cf. Steinbring, 1998d).

In using the epistemological triangle for analysing interactive mathematical knowledge, two central epistemological characteristics are questioned (cf. chapter 3):

- the manner of the constitution of the relation „object / reference context"<-> „sign / symbol",
- the temporary character and interactive exchange of the roles of „object / reference context" and „sign / symbol".

The above basic considerations regarding the epistemologically-oriented research approach lead to the following main aspects of its methodological procedures in the course of a qualitative analysis. The analysis is applied to documented episodes of mathematics teaching (or of interviews using mathematical communication, etc.). Such episodes are presented as transcribed (and mostly video-based) documentation. They are specially selected with regard to the main research question presented in this study: How is new mathematical knowledge interactively constructed in a typical instructional communication among students together with the teacher? This question is further elaborated in chapter 2.

Before an epistemological analysis of classroom episodes using the epistemological triangle is carried out it is helpful to first accomplish a „phenomenological" observation – a kind of „paraphrase" – of the interaction in the observed episode. The following steps of analysis are designed to serve this approach. This sequence is not to be used as an unchangeable and fixed recipe or as formal directives for use. It is more like a global perspective or heuristics which emerge during the work.

A. Descriptive Aspects of the Forms and Contents of Interaction

1. Types of Phases of Interaction with Regard to the Mathematical Content

- Introduction to *new* mathematical knowledge (the teacher introduces their students to a mathematical domain that they do not know yet)
- Exploration of new mathematical knowledge
- Students explain what they have worked on (homework, group work, etc.)
- Exercise, consolidating, work on old, already familiar mathematical knowledge (mathematical procedures are applied to exemplary exercises, exercises are calculated, etc.)

2. Organizational Forms of the Course of Interaction

- Formal instruction, the teacher works (in question-answer-form) with the whole class
- Interaction between the teacher and students working in small groups
- Silent work, or seat work (e.g. students solve problems by themselves)
- Pair work
- Group work

3. Description of the Mathematical Issue

- What mathematical domain is dealt with?
- What exercise or what problem is dealt with?

- By what means and in what form is the exercise or the problem dealt with?
- What solution is given?

*B. Linear **Time Structure** of the Transcript in Phases and Sub phases*

The given episode can usually be structured into a temporal course of single phases and sub-phases in a quite definite manner (cf. Mehan, 1979; Sinclair & Coulthard, 1975), in which the attempts of single students at solving the problem are accepted or refused by the teacher.

1. *Beginning of the phase*: The teacher (sometimes a student) poses a problem or proposes a new problem or a variant of an old problem.
2. *Interaction during the phase*: Students make suggestions or offer solutions.
3. *End of the phase*: The student's suggested solution is accepted, modified, or refused by the teacher.

The end of a phase is in most cases also *the beginning of the next phase*. One has to pay attention to the fact that this phase structure does not always proceed successively in the sense that only after the termination of a phase is the next phase opened; phases can also be interlocked (cf. Jungwirth, Steinbring, Voigt & Wollring, 1994).

*C. Self-organizing **Problem Structure***

The analysis tries to identify the changing types and forms of reasoning, explanation of solutions and answers within the phases with regard to the mathematical knowledge, such as, for example:

- Mathematical procedures, rules, „laws";
- The operative means used, graphs, diagrams, other visual representations, concrete contexts and real world situations;
- Reductive reasoning: reasoning which leads back to familiar (mathematical and concrete) issues,
- The way in which the relation between the new and the old knowledge is made.

The aspects and questions for a qualitative analysis stated under A), B) and C) include the *first step* of a „Description of the episode along the single phases" in the frame of the extensive epistemologically-oriented analyses of classroom episodes carried out in this work (cf. chapter 3). In this first step the instrument of analysis can be seen as „theoretical glasses", which allow for obtaining a first overview and ordering of the events observed in the teaching episode.

The epistemological analysis in its strict sense starts on the basis of this first step of analysis of the classroom episodes. Its central task is the attempt to reconstruct interactively created relations / structures between the „signs / symbols" and the „objects / reference contexts" in the temporal course of the mathematical communication. According to the analytic prerequisites of the epistemological triangle „the development of the epistemological interpretation in the course of the interaction" is accomplished in this *second step of analysis*. (cf. chapter 3). The two epistemological characteristics (1) of the manner of the constitution of a relation between „object

/ reference context“ and „sign / symbol“ and (2) of the temporal character and the interactive interchange of the roles of „object / reference context“ and „sign / symbol“ are the main focus of attention. This analysis step uses a kind of a „theoretical magnifying glass“ in order to classify the meaning and epistemological structure of the constructed knowledge more closely.

Further steps of analysis based on this second step pursue specific research questions in the frame of the *social epistemology of mathematical knowledge* approach. The *third step of analysis* is: „Detailed analysis of reasoning patterns from communicative and epistemological points of view“ (chapter 3). In this analysis different types of interactively constituted patterns of reasoning and generalizing mathematical knowledge are identified with the help of the available data. Krummheuer described and analyzed such patterns of reasoning in elementary mathematics instruction – he was presenting argumentation formats – from a sociological and communicative perspective (Krummheuer, 1997; 1998). In the research approach pursued here the interrelations between the epistemological characteristics of mathematical knowledge and the social conditions of the interaction when constructing new knowledge are in the center of attention. This analysis can be understood as the use of a „theoretical microscope“ which allows for searching for connections and reciprocal relations in the more indepth details.

In order to illustrate at least several core ideas of the epistemologically-oriented approach to interaction, I will present a short episode as an explainatory example in which the epistemological triangle is used as an important analytic instrument. The classroom episode was observed in grade three of an elementary school. With the teacher‘s assistance, children were developing their first insights into numbers beyond 100. The number space was to be extended from one hundred to one thousand. For this purpose the students would be using a variety of different means of visualization and structured diagrams, as, for instance, the number line, or the thousands book (Fig. 7) and the 1000 dots field. Based on their mathematical work in grade two, children were already used to working with such structured diagrams for interpreting and justifying arithmetical relationships and operations. During the mathematics lesson observed here, children were first of all expected to become familiar, to a certain extent, with the thousands book (Fig. 7).

In the course of the lesson, the teacher wanted the students to read numbers from the thousands book one after the other in steps of 50 starting with the number 50. Some of these numbers are written in this book in the familiar decimal notation; others are not written down, but have to be concluded structurally. This exercise, just like many others, is intended to support the children‘s‘ first orientation in reading and exploring this new number book.

During the first phase of the long episode the task proposed by the teacher is completed; one child after the other names a number in the series of numbers from 50 to 1000 in steps of 50. The teacher writes every number in mathematical notation at the blackboard. Two major problems arise in the beginning: „What is the next number after 50?“ Is it the 110, on the next page right on the top in the thousands book, or is it the 100? One student proposed, as the following numbers, „One hundred ten, two hundred.“ And the second problem: „Does 700 follow the number 200?“. Here, Julia goes on in steps of 500. These problems are interactively negotiated and clarified. Subsequently the children are able to quickly recite the numbers up to 1000.

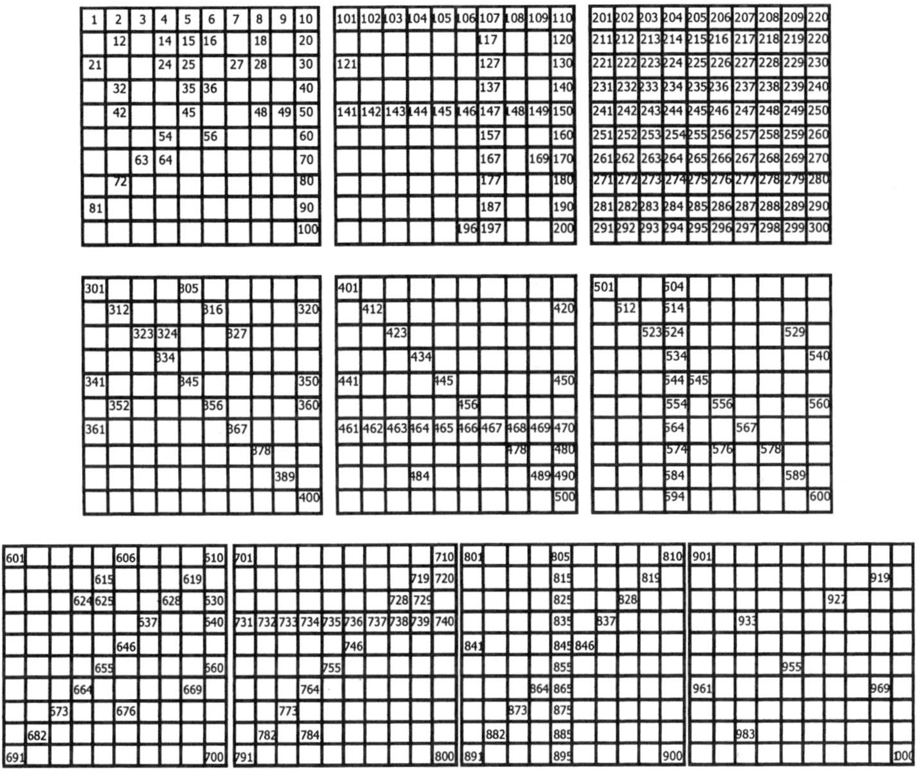

Figure 7. The thousands book.

61 S.: Seven hundred and fifty.
62 S.: Eight hundred.
63 S.: Eight hundred and fifty.
64 S.: Nine hundred.
65 T.: Nine hundred? And further?
66 S.: Nine hundred and fifty.
67 T.: Phillip!
68 Phillip: Nine hundred and fifty.
69 T.: Fifty, Marc?
70 Marc: One hundred, uhm, one thousand.
71 T.: Yes, and what comes then? After one thousand? Another fifty?

At this point a second phase starts. The emerging problem refers to questions such as, „What comes after 1000?", „What is this number called?", „What is the name of this number?", „How is this number written down correctly with decimal digits?".

After the number looked for has been correctly named by one student as „One thousand and fifty", this number is to be written down in decimal digits, i.e. as a mathematical symbol. Different proposals are made: Kai proposes the notation „1050", Marc writes down „1005" and Svenja writes „10050". Which of these proposals is correct?

The extension of the number space from 100 to 1000 that has been carried out at the beginning of the episode represents a construction of the new numbers in which these numbers are given by the concrete positions – and therefore are based on empirical qualities. The new numbers are for the time being quantities for positions in the thousands book. Accordingly, the mathematical notations 1050, 1005, and 10050 for the verbally named number „one thousand and fifty" represent something like abbreviations for the quantities. Since three different „mathematical names" for the number word „one thousand and fifty" are proposed, the question arises which mathematical name is the right one. It can be seen later that with this problem, the role of the mathematical number symbols begins to change.

In order to be able to decide which of the proposed notations is correct, the teacher now intends to refer to the position table as a means of providing justifications; the students are expected to look for the single positions of the new, big numbers.

111 T.: What hint could you give to Marc, Kai, and Svenja? We already have names for the different digits. Here we have a three digit number, and here we have a two digit number, here a four digit number, and there is also a five digit number. Uhm?

112 *Teacher points to the numbers on the blackboard: 950, 50, 1050, 1005, 10050.*

113 S.: And with five digits.

This implicit reference to the positions of the number might be a reason for Svenja to revise her notation. In this way, she obtains a correct representation; she says she has to change something and continues:

119 Svenja: Then I put there a one and a zero, and then over there the fifty.

121 T.: You think it is better this way? Why now?

122 *Teacher points to the 1050.*

123 Svenja: Yes, when I look up there.

124 T.: What did you look at? ... Everybody pay attention, please. What is, how is this number called?

125 Svenja: Two hundred and fifty.

126 T.: Uhm, and, how did this then help you here in this case?

127 Svenja: That I have to erase a number, because there, there are also only two hundred and fifty.

Svenja seems to make a comparison of the syntactic structure of numbers between the correct symbolic notation „250" for the number two hundred and fifty, and the imagined symbolic notation „2050" for this number according to the principle she has used for writing the number one thousand and fifty in mathematical terms as „10050". Consequently one thousand and fifty also could only be symbolically written as „1050". Svenja wants to erase a number (127).

Then the teacher explicitly introduces the position table after Felix's interpretation of the single digits as „units", „tens", „hundreds", and „thousands".

131 *Felix goes to the blackboard and points at the different positions of the number 1050.*

132 Felix: Actually, it's really simple, because this is the thousand unit, this the hundred unit, the tens, the

134 units.

The teacher writes the following table down on the blackboard, above the number 1050:

T	H	T	U
1	0	5	0

The question whether the notation „10050" for the number one thousand and fifty is correct cannot be answered on the basis of the prevailing empirical understanding of the new numbers as quantities of positions in the thousands book. Svenja looks at the internal structure of the digits of the number symbol by comparing it to the representation of the number two hundred and fifty which is already familiar to her. Felix refers to the characterization of digits by their place value designation: units, tens, hundreds, and thousands. With the help of Svenja and Felix's explanations, the interpretation of the symbols for numbers is beginning to change: These symbols are no longer a direct result of an empirical quality or abbreviated names for quantities, but numbers have their own internal structure and there is a production mechanism for numbers which makes them independent from the given reality.

According to these background considerations it is possible to make different interpretations of the epistemological triangle for the analysis of this teaching episode. In the beginning of the episode, numbers are understood as „existing" empirical properties in the thousands book and in this way children count in steps of fifty up to one thousand. One could describe this with the help of the following specification of the epistemological triangle: Numbers are names for empirical objects, for quantities, etc. and these names are verbal or mathematical. The numbers are verbally spoken out, written in their decimal digital representation and pointed at in the thousands book.

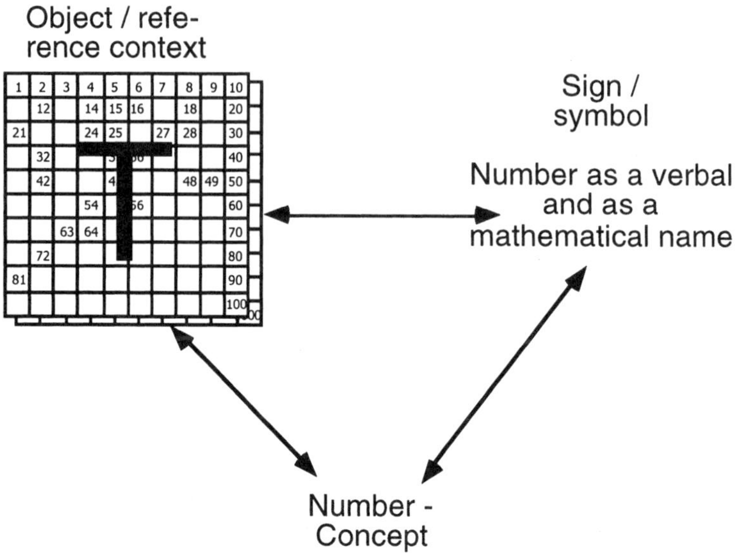

Figure 8. The epistemological triangle and the thousands book.

With the question: „Yes, and what comes after that? After one thousand?“ (71), the old context is exceeded and breaks down. Now the name „One thousand and fifty“ is to be written down in mathematical notation; the contributions and the critiques of the students clearly show how the verbal name and the mathematical name (the mathematical sign code) differ, and that they are no longer – in principle – identical. The „digital name“ and the „verbal name“ of the number are definitely different from each other!

On this basis there are now two different proposals made by children on how to decide which proposal is correct: Svenja corrects her number and justifies this by a comparison with the number 250. This comparison seems to be based on the syntactical structure of the digits, but it is not possible to follow Svenja‘s justification precisely. In this way, the interpretation in the epistemological triangle shifts again. Now the „object“ is no longer the thousands book but the syntactical structure of the digital representation.

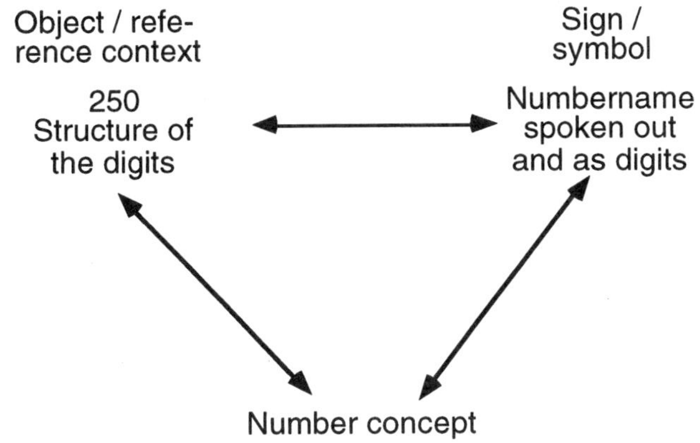

Figure 9. The epistemological triangle and the syntactic structure of numbers.

Then Felix remembers the interpretation of the separate digits in the mathematical name of a number as „Thousands, hundreds, tens and units" The teacher reinforces this by introducing the position table. Besides the syntactical structure of the number, this position table emphasizes the internal, systemic connections between the digits themselves. With this introduction of the position table (the discrimination between name and digit) the numbers obtain a new reference context, and once again, the reading of the epistemological triangle is changed.

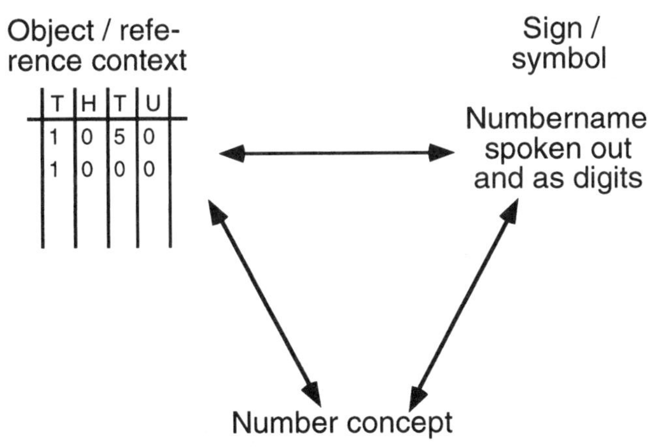

Figure 10. The epistemological triangle and the decimal structure of numbers.

Here, the reference context no longer consists of the thousands book and consequently of an empirical, given foundation on which, apparently, the new numbers can be built up. The new, changed reference context represents a symbolic, structured system itself. The numbers given by the mathematical symbols, as for example

1005 or 1050, are developed meaningfully by referring to internal structures present in the symbol itself – the position of digits and their relations to each other.

This small case study of using the epistemological triangle for analyzing a classroom episode of mathematical interaction shows, above all, how during the interaction, the interpretation of the number signs as names for an empirical object changes into a conception of symbols for arithmetic relations between the individual positions of the number signs. This way, the epistemological characterization of mathematical knowledge changes from an empirical interpretation to a relational view of central relations and structures. Further, the reference context is no longer understood as an empirical domain of objects and features, but interpreted as an embodiment of a structure, in this case as the decimal structure embodied in the decimal position table (cf. Steinbring, 2001).

Some Existing Research in the Frame of the Social Epistemology of Mathematical Knowledge Approach

Research carried out so far in the field of the epistemology oriented approach to mathematics classroom interaction for investigating structures between „sign / symbol“ and „object / reference context“ used qualitative methods as well as quantitative methods based on qualitative pre-analyses. A qualitative epistemological analysis of mathematical interactions generally begins with the identification of the interpretation of those elements of communication that are used by the participants and are explicitly related to mathematical knowledge. According to their use as „sign / symbol“ or as „object / reference context“ these communicative elements of those participating in the communication are classified. In a next step, the development over time, in the episode, of the structure of relations between „sign/symbol“ and „object/reference context“ is described.

Quantitative analyses that have been performed were based on the results of the above summarized general qualitative analyses. For instance, on the basis of using a detailed transcript for a number of mathematical lessons the mathematical contributions of the students and the teacher were separated into two groups. All mathematical statements of the participants have then been coded by two independent external coders as belonging to three categories: (1) „sign / symbol“, (2) „object / reference context“, and (3) expressing a relation between „sign / symbol“ and „object / reference context“. In a fourth category (4) those statements were listed that could not be classified in a strict manner.

These codings were „translated“ into graphic visualizations (with the help of a computer program) in the following way: The time development of the lesson was written on the x-axis by means of numbering all statements one by one (up to 300 statements). The three central categories were marked at the y-axis (in the form of a black bar for each). In this way the temporal course of the structure of the epistemological interaction could be made visible (similar to an electrocardiogram). This visualization of the structure of a lesson from an epistemological, socially constituted mathematical knowledge viewpoint made it possible for the researchers to gain qualitative insights into the type of the social generation of mathematical meaning in the class. If the students‘ and the teacher‘s statements were made graphically distinct, the visualization could highlight the contrast between the different types and allow interpretation of their influence on each other (Bromme & Steinbring, 1994).

In epistemological investigations of this kind (cf. Bromme & Steinbring, 1990a; 1990b; Bromme & Steinbring, 1994; Steinbring, 1993) the following two insights about the social construction of mathematical knowledge in everyday instruction have been gained. With regard to the course of interaction in mathematics instruction, it can be stated, in general, that there are two main types of construction of knowledge. One is a tendency to begin the development of meaning with concrete, empirical references for the symbols in order to carry out a more or less determined disconnection of the mathematical signs and operations from the concrete interpretations to an isolated, „abstract" level of interpretation. The other type, which may occur only sometimes in parts of a lesson, are characterized by leaving open the tension between interpretations of „sign / symbol" and „object / reference context". This is undertaken without closure which is too fast or definitive, to the benefit of a formal reduction of the mathematical signs to a fixed (empirical or formal) meaning. In classroom interactions of the latter type it was possible to observe a greater variability in the social interpretations of mathematical signs, symbols and aspects of concepts.

The „separation" of the students' statements from the teacher's statements, together with the corresponding visualization of epistemological structures, allows discussion of the influences on the two main types of construction of knowledge portrayed above. In classes where students followed the interaction-related routines quite strictly, there were no typical differences between the visual patterns of the students' and the teacher's statements. Here, mainly constructions of mathematical knowledge from empirical references to abstract, formal interpretations were carried out. On the other hand, where patterns of epistemological construction of mathematical knowledge of the students' statements differed from those of the teacher, a more open interpretation of the relation between „sign / symbol" and „object / reference context" was more likely to be obtained. Thus the observable interpretation of aspects of mathematical concept in the classroom communication was rather rich in relations.

A number of epistemology-oriented qualitative studies conducted up to the present have examined the interrelations between the communicative patterns and routines and the epistemological, socially-dependent knowledge constructions in everyday teaching. One research work elaborated the reciprocal influences between „communicative funnel patterns" (Bauersfeld, 1978) and epistemological circles of mathematical knowledge in mathematical interaction (Steinbring, 1997a). In another comparative study (Maier & Steinbring, 1998) a classroom episode related to the introduction of the concept of fractions in grade 5, served as a data basis to represent and compare two theoretical approaches to qualitative analysis of instruction, in which differences and similarities between an individual-cognitive and a socio-interactive interpretation of processes of mathematical meaning development became visible.

3. MATHEMATICAL KNOWLEDGE AND COMMUNICATION – COMMUNICATIVE SYSTEMS AS NECESSARY LIVING SPACES FOR PROCESSES OF MATHEMATICAL COGNITION

The subject of epistemologically-oriented analyses of interaction are, generally speaking, processes of communication about mathematics. In the preceding sections it has been stated, among other things, that mathematical knowledge in the contexts of its acquisition should be understood as a dynamic process of mathematization. Thereupon, the particular epistemological characteristic of mathematical knowledge consists in the (mostly invisible) structures and relations which are coded by means of signs / symbols and related to reference contexts. From this point of view, mathematical knowledge can be conceived of as still being open and not fixed in learning and acquisition processes. Also when the learner interprets mathematical signs on their own and in a varying manner, one can indeed finally speak of a personal mathematical communication. If one assumes on an always definite, and fixed given body of mathematical knowledge, there could not be open, lively mathematical communication. Mathematical communication would be reduced to a direct exchange of valid and definite mathematical facts.

In this research project, processes of mathematical acquisition and understanding are seen as processes of interpreting signs with regard to a corresponding reference context. On the basis of such a premise of mathematical knowledge and its recognition it is necessary, to develop a suitable understanding of the concept of communication, specific to the communication content in question, namely, mathematics. With respect to teaching-learning-processes one could proceed on the assumption that the role of communication is completely different and isolated from the role of the mathematical content. Thus one could analyze the communicative structure itself, without considering the influences of the contents of instruction. At the other extreme, one could assume that mathematical ideas and contents can be exchanged directly between the participants in the communication, and therefore mathematical knowledge is already completely included in the communicative actions. Consequently there would be no partial autonomy of mathematical knowledge or of communication.

The problem of the completeness of relations between processes of interaction or communication and mathematical knowledge depends strongly on the premises regarding the epistemological nature of mathematical knowledge and the concept of communication. In the preceding chapter some suitable and useful premises for this purpose have been developed with regard to the epistemological interpretation of mathematical knowledge in classroom interaction. Accordingly I will now try to develop an understanding of the communication concept which „matches“ this conception of the mathematical content. In order to do this, I will consult, in particular, the concept of communication elaborated by the German sociologist Niklas Luhmann, which he furnished in the foundations of his system-oriented characterization of society together with its social sub-systems (Luhmann, 1997).

The interplay between the forms of communication in the course of teaching and the epistemological interpretations of mathematical knowledge created in the interaction (implicitly or explicitly) is a central theoretical component of the epistemologically-oriented interaction analysis. The reciprocal actions between the central elements, „forms of classroom communication“, „interactively constituted interpretation of mathematics“ and „nature of mathematical knowledge“ will be explained by presenting an illustrative example., that is by means of an analysis of a short teaching episode.

In a 6th grade class different descriptions of outcomes of rolling a die are noted as events, e.g. „Throw a 6!“, „Throw a number smaller than or equal to 2!“, „Throw an even number!“ etc. On the black board events such as {6}, {4}, {1, 2}, {2, 4, 6}, {3}, {5, 6} or {1, 2, 3, 4, 5, 6} are noted.

At this point, the episode „Is the impossible event possible?“ begins.

1	T	There is a subset that we will never write, but we want to mention it.
2	S	Open brackets one, three, five close brackets.
3	S	...the odd numbers.
4	T	Yes, we did that.
5	S	Number smaller than or equal to one.
6	T	One could do that. How would you write that as a set?
7	S	One in parentheses.
8	T	Right! If I said now: Throw a number smaller than one?
9	S	...that does not work!...
10	T	That is also an event! this event...
11	SS	...doesn‘t work! ...doesn‘t work!!
12	T	How could we provide that with an adjective?
13	S	...the uncertain event.
14	T	We simply say the impossible event. What kind of subset is that, when I speak of the impossible event?
15	S	That‘s the empty set, nothing in brackets.
16	S	All that is not true!

From a point of view that concentrates mainly on the communicative contributions in this episode, one could deduce that several suggestions by the students follow the teacher‘s prompt (1): the set or the event of odd numbers is offered (2,3); the teacher indirectly signalizes rejection (4), and as a consequence the event „number $\leq$ 1“ is offered and apparently accepted by the teacher with the question as to how it is written as a set. The answer – {1} – is accepted, the student‘s suggestion is taken up and modified by the teacher: „Throw a number smaller than one!“ There are contrary reactions: „That doesn‘t work!“. However, the teacher explains this description as a (possible) event in the question as to what property can this event be described with? In meaning negotiations with the students the impossible event is introduced by the teacher on the basis of one student‘s proposal to speak of an uncertain event here. Also there is the question of which set belongs to this impossible event? One student seems to remember the empty set (15) while another students reacts with a denial: „All that is not true!“ (16).

An analysis focused exclusively on communication alone could, for example, explore the „logic“ in the sequence of the contributions and try to explain how certain statements might have occurred. How is, for instance, the suggestion „...the uncertain event“ (13) to be interpreted? The teacher had characterized the search for an event as an extreme one just like the „certain event“ discussed before. This comparison might have caused the student to speak just of an uncertain event.. Furthermore the teacher takes up the student‘s suggestion „number $\leq$ 1“ and modifies it to „number < 1“, and this is certainly remarkable for an analysis from a communication point of view. In addition, statements such as the last one: „All that is not true!“ (16) can be interpreted in many different ways: Impossible events are impossible, empty sets are no sets, or all that is rubbish. Even according to such elementary goals of mathematical instruction a purely communicative analysis of mathematical classroom interaction does not say anything about the mathematical quality or increase in children‘s learning.

Starting from the assumption that mathematical terminology describes mathematical objects in an unequivocal way, this classroom interaction – with the familiar patterns and routines – leads to a correct, useful mathematical goal: the impossible event corresponds to the empty set. At least one student seems to remember this, and perhaps it also stimulates the interest of some of his classmates. From this perspective, a purely communicative analysis might supply additional indications as to how mathematical interactions may proceed, where the aim is to convey a body of fixed, definite mathematical knowledge.

However, according to the basic assumption made here that mathematical knowledge can only emerge in the interaction (and in the activity), the learning and understanding of mathematics is a process of interactive interpretation of mathematical concepts and notations and not a process of conveying already existing ready made knowledge to students. With this view in mind, a specifically communicative analysis should be carefully complemented by an epistemologically-oriented analysis. In the course of this teaching episode the students are asked to describe events of the game with mathematical notations and signs: The rule of the game „Throw an even number!“ corresponds to the event {2, 4, 6},and this mathematical notation is identified with the corresponding set or subset. In this way the interaction is about the interpretation of new signs or symbols in the context of a real experiment with throwing a die in the frame of certain rules. This makes the key point of an epistemological perspective: How do the new mathematical signs and symbols acquire meaning and of what type is this meaning?

The examples of „Throw a 6!“; „Throw a number smaller than or equal to 2!“ etc. correspond to concrete, practicable die experiments with observable outcomes. Already the „extension“ of the rules of the game to „all numbers“ and then to the even stronger example of „Throw a number $\leq$ 1!“ goes beyond the concrete game context. These „unrealistic“ rules do not offer the same justifying context for the corresponding mathematical signs as the examples treated before. They are indeed impossible, in an immediate sense. The introduction of the „impossible event“ and the „empty set“ require and cause a fundamental change in the justifying context for these concepts. These concepts are only understandable as relational elements of a structural, operative net, and no longer as empirically explainable mathematical notations (this is comparable, e.g., to the transition from the positive numbers to zero, or later to the negative numbers).

Under an epistemologically oriented analysis a conflicting interpretation for the explanation of the concept „event" (the interpretation of the mathematical signs) in the transition from an empirical interpretation to a necessary – but not realized – relational interpretation can be stated in the present episode. Several students seemingly want to express this „impossibility" of creating mathematical meaning in the empirical justification at hand with statements such as „...that doesn't work!" (9,11). One student then engages in the familiar interaction ritual with the teacher and the „impossible event" and the „empty set" are routinely generated on the communicative level. But at least one other student can resist this seemingly bad compromise of a commonly negotiated knowledge interpretation with the words: „All that is not true!" (16).

In the following, essential aspects of the communication concept according to Luhmann will be presented. The problem of the difference and the demarcation line between the social and the individual (cf. section 1.2) is discussed in this context again. Teaching and learning contains the two fundamental dimensions of the interactive-social and the cognitive-individual side of processes of knowledge development. In the frame of system-theoretical analyses Luhmann (cf. among others Luhmann, 1996; 1997a) elaborated the difference and the fundamental separation of social and psychic processes. Furthermore, the autopoietic functioning of communication or interaction is described in a manner which focuses on the construction of the interactive constitution of relations between „signs / symbols" and „objects / reference contexts", formulated in this theoretical approach, as the central characteristic of mathematical communication.

The difference between psychic and social processes is explained by Luhmann as follows:

> Pedagogy will hardly admit that psychic processes and social processes operate in completely separate manners. But the individuals' consciousness cannot reach other individuals with its own operations. It can perceive, it can perceive being perceived and can try to influence its own body in such a way that no damage results from being perceived. It can be seen and heard and control the image to a certain extent. But when communication is to come about, a completely different, also closed, also autopoietic system has to become active, as a social system which reproduces communication by communication and does not do anything else than this (Luhmann, 1996, p. 279).

The concept of the „autopoietic system" has been introduced by Maturana and Varela (cf. for example Maturana & Varela, 1987). It describes self-referential systems, which mean living systems which exist and develop autonomously and which produce and re-produce those elements that are needed for their existence in order to maintain them. These systems produce the necessary elements in the system itself without any external intervention. Not only biological processes are examined within the concept of the „autopoietic system", it is also applied to social and psychic processes. Luhmann has extended the concept of autopoietic systems to systems within society. The essential property of such non-biological systems is a specific kind of operation that only takes place and is re-produced in this system. The main non-biological systems in society are the social and the psychic system. What is the core difference between a social and a psychic system? Psychic systems are based on consciousness and social systems are based on communication.

> A social system cannot think, a psychic system cannot communicate. Causally seen, there still are immense, highly complex interdependencies (Luhmann, 1997b, p. 28).

This theoretical approach to the relation between the social and the individual has consequence for mathematics education. There are no simple, direct dependencies or immediate influences between the social system of mathematics instruction with its communication structures and the psychic system of the students' learning mathematics. The influences of the instruction on the students' learning are complex interdependencies. Thus instruction cannot automatically and directly induce understanding in the children's consciousness.

> Understanding is never a mere duplication of the message in another consciousness, but in the communication system itself the prerequisite for further communication, thus a condition of the autopoiesis of the social system. Whatever the participants in their respective own self-referential closed consciousness may think of it: the communication system elaborates its own understanding or misunderstanding and for this purpose creates processes of self-observation and self-control (Luhmann, 1997b, p. 22).

But what then are the possibilities and conditions for the communication system „teaching" in order to stimulate the student's psychic system „consciousness"? How is communication adapted to „... this will-o'-the-wisp of a consciousness" (Luhmann, 1997b, p. 30)?

> Communication systems and psychic systems (or consciousness) form two clearly separated autopoietic domains; ... These two kinds of systems are, however, connected to each other in a particularly tight relation and mutually form a ‚portion of necessary environment': Without the participation of consciousness systems there is no communication, and without the participation in communication there is no development of the consciousness (Baraldi, Corsi & Esposito, 1997, p. 86).

Language is the central medium for the creation of possible „connections" between communication and consciousness.

> The specificity of the relation of communication and consciousness is ... connected to the fact that this coincidence does not happen by chance thanks to the availability of language ..., but that it can be expected and partly planned. ... [one] can ... say that verbal communication can treat psychic systems as a medium which is always prepared to take on communicative forms. Consciousness on the other hand can use language in order to treat communication as a medium into which it can forge its shapes; since a verbally expressed thought can always be communicated and thus force the communication process to work up a psychic stimulus (Baraldi, Corsi & Esposito, 1997, p. 87).

In Luhmann's system-theoretical description, „communication" is the fundamental concept of sociology. „Communication is the last element or the specific operation ... of social systems. It consists of the synthesis of three selections: (1) message; (2) information and (3) understanding of the difference between information and message" (Baraldi, Corsi & Esposito, 1997, p. 89). With this theoretical basic perspective Luhmann describes society with its social sub systems as a whole. I will only deal with the central „mechanism" of communication as far as it is relevant for the research approach presented in this work.

> One talks of communication when Ego understands that Alter has communicated a message; this information can then be ascribed to him. The communication of an infor-

> mation (Alter for instance says „It's raining") is not yet a real information. Communication only realizes itself if it is understood: if the information („It's raining") and Alter's intention for the message (for example, Alter wants Ego to take an umbrella with him) are understood as distinct selections. Without understanding communication cannot be observed: Alter waves his hand to Ego, and Ego continues walking because he has not understood that the waving was a greeting. Understanding realizes the fundamental distinction in communication: the distinction between message and information (Baraldi, Corsi & Esposito, 1997, p. 89).

Examples for a communication understood in such a way can be found in the classroom episode mentioned above. For instance, Alter – the teacher – says: „Yes, we did that" (4). And Ego – a student – shows by means of his message that understanding was observable: a „Number smaller than or equal to one" (5). With his message the teacher did not only want to give the information that this example had appeared once before, but that the case intended by him has not been stated yet; the student tries another example, and he adds a further message to this. The communication at the end of the episode could be interpreted in the following way. The teacher formes the message: „The impossible event is possible.", in which he seems to have the intention of formally equating the (not realizable) game rule „Throw a number < 1" with the other rules „Throw a number ≤ 1" or „Throw a number > 1". A student states: „The impossible event is impossible" and connects a piece of information other than that intended by the teacher to this message; while another student seemingly makes an understanding observable by his message „The impossible event is the empty set.".

With language as a means of connection between communication and consciousness one has to distinguish between sound and sense while in written language one accordingly distinguishes between sign (or signifier) and sense. This difference between sign and intended meaning is the starting point of the autopoiesis of the communicative system.

For the elaboration of the conditions of the autopoiesis of the communicative system Luhmann refers among others to the works of de Saussure who distinguishes between „signifier (signifiant)" and „signified (signifié)", and these two sides constitute the sign. „Besides the concepts signifier and signified Saussure also uses the concept of sign: The *sign* [*signe*] designates the whole, that which contains the signifier and the signified as its two parts" (Nöth, 2000, p. 74).

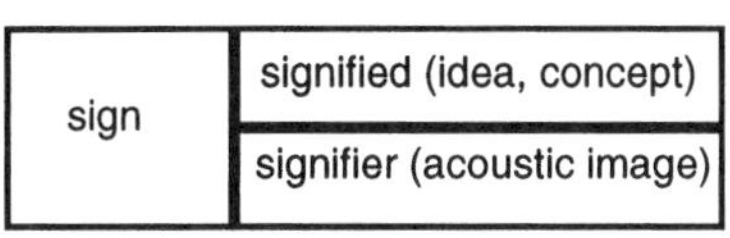

Figure 11. The dyadic model of de Saussure's sign.

De Saussure gives the following – often cited in the literature – example to explain his interpretation of the verbal sign.

> The verbal sign thus is something actually existing in the mind, which has two sides and can be represented by means of the following figure:

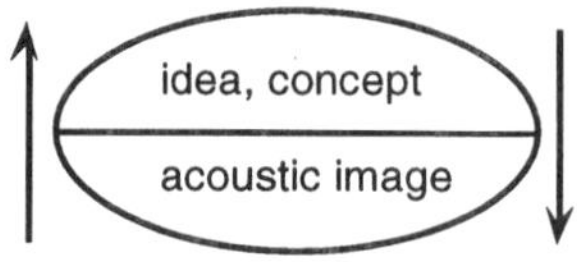

Figure 12. The dyadic model of de Saussure's sign.

> These two components are closely tied to each other and correspond to each other. Whether we are looking for the sense of the Latin word *arbor* or for the word which in Latin designates the conception „tree", it is clear that only the assignments valid in this language seem appropriate, and we exclude any other assignment one could think of.

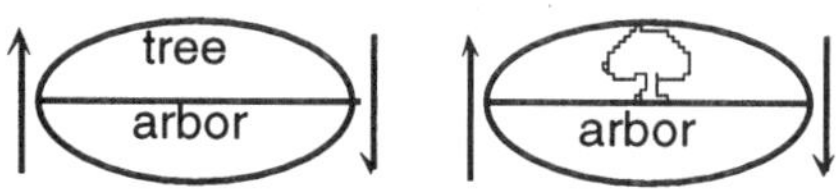

Figure 13. The dyadic model of de Saussure's sign with the example of „tree".

> With this definition an important terminological question is raised. I call the connection of the concept with the acoustic image the sign; yet according to the common use this term designated the acoustic image alone, e.g. a word (*arbor* etc.). One forgets that, if *arbor* is called sign, this only counts in so far as it is bearer of the conception „tree", so that this designation includes not only the thought of the sensorial part, but also the thought of the whole (de Saussure, 1967, p. 78).

Luhmann uses this conception of the sign as a fundamental element in his theory of communication in order to explain how the autopoiesis of the communicative system is kept going; he writes:

> Signs are also forms, that means marked distinctions. They distinguish, according to Saussure, the signifier (signifiant) from the signified (signifié). In the form of the sign, that is in the relation of signifier to signified, there are references: The signifier signifies the signified. The form itself (and only this should be called sign [31]), on the other hand, has no reference; it only functions as a distinction, and only then when it is factually used as one. ...
>
> [Footnote: 31) In German it is hard to keep this up from a linguistic aesthetical point of view, and thus repeated confusions of signifier and signified occur in the related literature] (Luhmann, 1977a, p. 208f.).

How is the autopoiesis of the social, thus of communication, possible? According to Luhmann the participants in the communication or in the communicative system reciprocally furnish „signifiers" (references) by means of *messages* (or actions) that point to *information* or „signifieds".

> It might ... be decisive that speaking (and gestures imitating it) illustrates one of the speaker's intentions, thus forces a distinction of information and message and in the following then a reaction to this difference with as well verbal means (Luhmann, 1997a, p. 85).

The messenger can only contribute a signifier, but the signified intended by the messenger, which can only lead to an understood sign, remains open and relatively indeterminate, and it can only be created by the receiver of the message by articulating a new signifier. Luhmann explains this in the following way:

> We do not start from the speech action which only occurs when one can expect it to be expected and understood, but from the situation of the receiver of the message who observes the messenger and ascribes the message, *but not the information* to him. The receiver of the message has to observe the message as the designation of an information, thus both together as a sign (as a form of distinction between signifier and signified)... (Luhmann, 1997a, p. 210; emphasis by the author).

The receiver is *not* allowed to assign the possible signified *strictly* to the speaker, but they have to „create it themselves“, as it emerges in the social communication.

The possible detachment of the information belonging to the message from the messenger is the starting point of the autopoiesis of the communicative system:

> This increases the possibilities of exposing oneself to certain environments or of escaping them, and offers the participants‘ self-organization the chance of distancing themselves from what is communicated. One remains perceivable, seizable, but only in that which one contributes to the communication in a well thought out way. This has the consequence that, with the normalization and recursive strengthening of these coupling operations, an autonomous autopoietic system of verbal communication emerges which operates self-determinedly and at the same time is fully compatible with the reflected participation of individuals. Now a co-evolution of individuals and society occurs, which over determines possible co-evolutive relations between individuals (for instance mother-child-relationships) (Luhmann, 1997a, p. 211).

This theoretical foundation of social communication as an autonomous system serves the epistemologically oriented research approach as a general basis of the designation of social, mathematical interactions as autonomous, self-organizing processes. This „mechanism“ explained by Luhmann keeps the autopoiesis of the communicative system running as the participants are permanently exchanging „signifiers“ which become socially constituted, communicative „signs“ by means of the presentation of new „signifiers“ by other participants. This mechanism describes a general way of the functioning of communication on whose foundation the special mathematical interaction takes place and can be analyzed at the same time.

The difference, decisive for the functioning of communication as an autopoietic system, between „signifier (signifiant)“ and „signified (signifié)“, which constitutes the „sign (signe)“, corresponds, in mathematical classroom interaction, to the difference in the epistemological triangle between „sign / symbol“ and „object / reference context“. The mediation between „sign / symbol“ and „object / reference context“ is regulated by means of the mathematical concept or conceptual aspects during the process of the epistemological construction of mathematical knowledge by the subject or the student. In a first approach the interrelations between „sign / symbol“ and „object / reference context“ in the epistemological triangle can be grasped through an analogy with the semiotic model (according to de Saussure); but the epistemological triangle also contains particularities with special reference to mathematical communication. Possible equivalents and specifics of the epistemological triangle compared to the semiotic model are discussed in section 2.2.

The system-theoretical explication of communication represented by Luhmann as a „lively“, continued, dynamic exchange of messages in which the observer can notice the constitution of signs, offers possibilities of productive connection to the epistemologically oriented approach. The just mentioned correspondence between

„signifier“ and „signified“ in the semiotic model and the distinction between „sign / symbol“ and „object / reference context“ in the epistemological triangle represents a complementarity which makes it possible to both understand mathematical communication as vivid, autopoietic systems with a dynamic development, and simultaneously analyzing and assessing the constructions of mathematical knowledge occurring during the interaction with the help of the epistemological analysis and with regard to their epistemological status.

Summing up, the research approach of the *social epistemology of mathematical knowledge in classroom interaction* is based on the three following, central components:

- The social epistemology of mathematical knowledge understands the „way of existence“ of mathematical knowledge and mathematical concepts as „social facts“ and the creation of social mathematical facts – particularly by learners – is subject to the interrelationships between the social construction and the epistemological necessity of mathematical knowledge.
- The social interaction theory of teaching and learning mathematics – where the social epistemology belongs in the widest sense – understands everyday mathematics instruction as a culture of its own with special symbols and cultural rituals.
- Mathematical communication can be regarded as an autopoietic system (according to Luhmann) where the autopoiesis of the system „mathematical communication“ is kept going by means of the relation „sign <-> reference context“ (generally: „signifier <-> signified“).

OVERVIEW OF THE SECOND CHAPTER

In the first chapter, the following fundamental positions of the empirical research have been elaborated as the basis of the empirical analysis. The development of (new) mathematical knowledge always takes place in the frame of social contexts or cultures. This is true for the socio-historical development of mathematics and the actual professional mathematical research practice as well as for the practice of teaching and learning of mathematics in instruction.

With the embedding of the development of mathematical knowledge into social environments, two central perspectives emerge: the epistemological characterization of mathematical knowledge and the communication as well as interactive development of the meaning of mathematical knowledge. The mathematical sign / symbol represents a central connection between these two perspectives. The mathematical sign / symbol is used for the „codification" and the epistemological characterization of mathematical knowledge and it is at the same time a fundamental element of any mathematical communication.

The epistemological triangle is used for the epistemological analysis of mathematical knowledge and mathematical interactions. The epistemological triangle can be taken as a conceptual scheme, which helps provide a connection to the general communication or discourse analysis can be produced, at the same time paying attention to the particularities of mathematical knowledge.

In the second chapter, new theoretical views and especially the research question of this study follows these fundamental positions.

The fundamental viewpoint taken in this whole work understands mathematical knowledge not as a mere collection of finished knowledge components, i.e. not as a repertoire of fact knowledge, which slowly expands. Mathematical knowledge in its further development is permanently subject to processes of new interpretation of the present knowledge, and these processes are especially determined by generalizations and abstractions. Thus new mathematical knowledge cannot be seen as elements, which are simply added to the familiar knowledge. Instead, true new mathematical knowledge always requires the construction of a new, generalized relation already in the present knowledge, thus a new view of the familiar knowledge. This difficult relation between old and new mathematical knowledge is discussed at the beginning of chapter 2 and in section 2.1.

The relation between old and new knowledge is essential for the epistemological characterization of mathematics and also plays an important role for the analysis of interactive construction processes. If, however, mathematical knowledge is understood as a finished and given product, which can be described by means of mathematical signs and symbols in an unequivocal way, then the development of mathematical knowledge and the learning of mathematics are simply different combinations of the signs and symbols, which then describe the progressively expanded knowledge. If mathematical knowledge is thus „defined" by means of unequivocal sign representations and if new mathematical knowledge emerges simply out of the combination of present sign representations, then teaching and learning mathematics can only be understood as a handing-over of the fixed knowledge elements from the teacher to the students.

The construction of new mathematical knowledge is a process of continued abstractions and generalizations in history as well as in actual professional research practice. When developing new mathematical knowledge for the students, teaching and learning mathematics also faces the problem of how the children's learning activities can be stimulated in such a way that not only further fixed fact knowledge is gathered, but that generalizations and new interpretations of the already present mathematical knowledge can be actively carried out.

For the use of the mathematical signs and symbols, the epistemological triangle – with the idea of „communication" as an autopoietic system in the background – opens a certain freedom and vagueness in the subjective explanation of these signs with regard to the accompanying reference contexts. The meaning of the mathematical signs and symbols is thus not determined beforehand, but it emerges in the production of a relation to a reference context and the type of this relation is also developed in the interaction.

The interpretation of mathematical signs and symbols can be relatively open and can be made in different ways. The learners, in their construction processes, are requested to actively carry out their own interpretations of the signs and symbols. At this point, it is essential for the students to succeed in producing useful interpretations of the new and unfamiliar mathematical signs and symbols using their abilities and in their own words. Otherwise, the communication produced out of the expectations of teaching and learning mathematics can lead to an interaction process in which an „immediate" transfer of the mathematical knowledge and of the interpretation of mathematical signs is carried out between teacher and students. This problem is discussed in section 2.2.

Teaching and learning mathematics are subject to the expectations and target perspectives of conveying and understanding a certain subject matter, here of mathematics. Thus the teaching communication is often influenced and steered by the challenge of the instruction process. The participants in a teaching communication often adjust their contributions and statements to the expectations and goals inherent in teaching, i.e. to learn the knowledge in question and to acquire the correct mathematical idea. In order to do this, the communicative contributions of teacher and students are often coordinated in such a way that they reciprocally support and complete each other so that in this way the goals of the communication are fulfilled more or less „automatically", for example to grasp correct mathematical statements.

Section 2.2 elaborates on the concept that the epistemological and the communicative perspective on mathematical knowledge in interactive situations complete each other. Accordingly, the interpretation of the signs / symbols in the epistemological triangle is put into a comparable relation to the communicative interpretations of the signifiers in the frame of communication as an autopoietic system. In the interweaving between the epistemological conditions of mathematical knowledge and the instructional communication, it is especially important to sensitively observe whether a true mathematical communication takes place in the instruction, in which the students are equal partners in the common, active construction of mathematical meanings and new knowledge. If mathematical knowledge is understood as finished, given fact knowledge, then the instruction communication often degenerates to a stream of social events which „define" each other reciprocally, and the students do not construct mathematical knowledge and interpret mathematical signs or symbols in autonomous, active participation. In this frame, the instructional communication simply serves for skillfully asking for the present knowledge familiar to the teacher and in this way formally fulfilling the demand for delivering and learning mathematics.

The theoretical question examined in this work aims at a careful and sensitive description of the details of teaching interactions, in which the participants mathematically communicate to each other in a true manner. These teaching interactions are about mathematical knowledge, which is here understood in a specific way as a consistent structure full of relations, which is generalized and re-interpreted in the further construction processes in the course of teaching. Besides, it must be taken into account that the instructional communication process is subject to its own conditions, which partly harmonize with the epistemological features of mathematical knowledge and its construction, but can partly be in conflict.

The relation of the two dimensions „Epistemological characterization of the mathematical knowledge" and „Communicative interpretation of the mathematical knowledge" forms the fundamental scheme for an appropriate analysis of the true mathematical communication and construction processes in teaching to be described. Accordingly, in section 2.3, on the basis of the clarification of the theoretical positions on mathematical knowledge and communication, a grid with these two dimensions is developed.

The analysis grid is supposed to make the classification of instruction episodes possible. The basic theoretical positions are incorporated into the construction of the grid in the following way. The relation from old to new knowledge is discussed in the tension between fact knowledge and relational or structural knowledge along the „communicative dimension". In this dimension, interactive knowledge constructions are evaluated according to whether they mainly treat fact knowledge, or whether they express relations in the knowledge. It is also evaluated as to how far a balance of the potentially new knowledge relation to the already present old knowledge is produced.

The epistemological dimension contrasts the two following views about (school) mathematical knowledge. On the one hand there is an empirical, situated view of the knowledge, which is concrete and embodied in examples; on the other hand there is a relational generality of the knowledge, which is contained in intended relations and structures, which exceed the concrete examples. Besides, there is a distribution according to these two views, attention is again paid to whether a balance between the situatedness and generality of the knowledge is produced in the interactive constructions for the knowledge.

With this analysis grid, especially such interactive knowledge constructions are to be classified as successful mathematical communications, in which the balance between old and new mathematical knowledge, and between situatedness and generality of the mathematical knowledge can be realized on the communicative as well as on the epistemological dimension.

In the first two chapters, the essential theoretical foundations, referring to the views about the nature of mathematical knowledge and about the role and meaning of the teaching communication taken here, have then been elaborated. Then the empirical investigation with the detailed qualitative analyses of selected instruction episodes can build on these.

CHAPTER 2

THE THEORETICAL RESEARCH QUESTION

In this research work mathematical classroom communication is analyzed from an epistemological perspective and using qualitative methods. The question at the center of the research interest is: How is new mathematical knowledge interactively constructed in a typical instructional communication among students together with the teacher?

With this question the emphasis is put on „*new* mathematical knowledge". Thus the aspect of the new knowledge as opposed to old knowledge is given special attention and a fundamental conceptual distinction between old and new knowledge is made. This problem is clearly linked to the famous „learning paradox" that will be discussed later.

In a first approach to the conceptual problem of the relation of old to new (mathematical) knowledge let us assume the following characterization. If the – still unfamiliar, thus potentially new – mathematical knowledge has been elaborated correctly according to the criteria of scientific demands, then it has to be logically and consistently connected with the already existing knowledge, and it would even have to be derivable from the existing knowledge. But if knowledge has been obtained this way as a logical deduction then it is not really new, since it follows on from the old knowledge. The sociologist Max Miller formulated this essential aspect in the characterization of the difference between old and new knowledge in the following way:

> Any learning or developmental theory can ... legitimately be expected to give an answer to the question how the new can emerge in the development. ... Every answer to the question ... is... subject to the following validity criterion: it has to show that the new in the development presupposes the old in the development and still systematically exceeds it, otherwise there can be no new or the new is already old, and the terms ‚learning' and ‚development' become meaningless (Miller, 1986, p. 18).

This criterion of distinction of actual new from already existing old knowledge is developed by Miller in the frame of his theory of learning as collective argumentation (Miller, 1986). In this present study, however, this distinction is at the same time essential for the development of mathematical knowledge in discovery or research contexts. New mathematical knowledge is thus not understood as knowledge logically deduced from already existing knowledge, it has to exceed the old knowledge systematically in order to be considered new knowledge.

For mathematical communication and classroom interaction as analyzed in our research, the problem between old and new knowledge leads to the following research questions. Can new mathematical knowledge emerge in the scientific research context, where all accepted and scientifically correct mathematical knowledge has to be consistently connected to the already existing knowledge as a matter of fact? How can the problem of the relation between new and old knowledge be adequately understood in the context of mathematical research? Analogous questions arise in the context of learning: Can insights into really new knowledge occur in the learning context at all? If learning is seen as a cognitive process of the individual, which is built upon the circumstances given and is dependent on the learner's cognitive abilities, then additional fact knowledge can result, but not really new knowledge systematically exceeding the old knowledge.

Let me mention at this point that, in this work, learning is not regarded as restricted to individual processes proceeding in the single individual, but it is perceived essentially as a process of active construction of knowledge by the learners in interactive instructional settings. Thus learning in all its possible appearances is no passive intake of knowledge, but ultimately always an active construction by the learner in the interactive discussion with others.

Miller (1986) emphasizes that in those „learning processes" that are restricted to the single learner and their cognition, the (young) students can ultimately only add knowledge derived from their already present knowledge, which is therefore not really new knowledge. For explaining the fact that individuals are also able to really learn and construct actual new knowledge by themselves he introduces the concept of „autonomous" learning. Autonomous learning is a kind of pseudo social interaction of the (experienced) learner with himself or herself, in which the learner can put critical questions to themselves and is able to make „contradictions" to their former assumptions about the knowledge in question.

From the point of view that learning is an individual, cognitive acquisition of knowledge, the problematic relation between old and new knowledge has been presented as the „Learning Paradox".

> The paradox is that if one tries to account for learning by mental actions carried out by the learner, then it is necessary to attribute to the learner a prior cognitive structure that is as advanced or complex as the one to be acquired (Bereiter, 1985, p. 202; cf. also Hoffmann 2000).

The paradox is based on the premise that any knowledge representation is only a combination of existing representations and „genuine" learning does not occur from this point of view.

> ... learning and development are incapable of creating emergent representations; they can only create new combinations of old representations. ... neither learning nor development, as currently understood, can construct emergent representation; therefore the basic representational atoms must be already present genetically (Bickhard, 1991, p. 16).

Interaction theory in mathematics education, on the other hand, emphasizes the „emerging nature of representations" according to which „interactive representations" emerge as functional relations of the system with its environment.

> Interactive representation is a matter of a system's functional relationship with its environment, not a matter of structural relationship. Any notion of the environment creating a structural relationship, therefore, is irrelevant. Furthermore, there is no possibility that the environment can passively impress or create a system. Interactive representation, then, can be created only by the internal constructions of the system itself. The system must try out new system organizations and test them against the environment, a variation and selection constructivism (Bickhard, 1991, p. 24f).

Luhmann's analysis of communication as an autopoietic system that is separated from the psychic system but needs it as a stimulating environment can be seen as a proposal of a solution to the „Learning Paradox". The relation between the social communication and the human's cognitive acquisition is interpreted in a new way. This implies that the relation between the interactive mathematical knowledge constructions and the individual conscious mathematical learning process is seen from a new perspective too. In communication systems new signs are interactively constructed by the mediation of signifiers. And in the process of „interpenetration" (Luhmann) these new signs can indirectly influence the learner's consciousness.

> The features of verbally formulated communication lead to the fact that it almost inevitably ‚fascinates' the present consciousness systems and drags them into its own processes; when a conversation is held nearby, it is almost impossible to pursue one's own train of thought unimpressed by that. ... Language thus creates the structural coupling of two systems which, even if they always remain separated, can expect reciprocal participation in the constitution of their system complexity. In this way – by repeating and expecting the convergence – a co-evolution of the interpenetrating systems emerges (Baraldi, Corsi & Esposito, 1997, p. 85ff).

The contrast between the individual and the social, or the question of the *priority* of the individual over the social or of the social over the individual is thus interpreted from an alternative position, namely as a mutual relation between individual consciousness and social communication. By means of socially created signs – and thus by means of social construction of mathematical knowledge – the learner is inspired to construct new cognitive structures in his consciousness.

Furthermore, for the purposes of this research focused on mathematical knowledge and its interactive construction pursued here – thus for the study of specific *mathematical* interaction processes – it is important to explain the specific requirements for new mathematical knowledge that is linked to old knowledge and at the same time must systematically exceed the old. A construction of new knowledge can be described initially as follows. In the domain of the existing mathematical knowledge signs / symbols have a more or less accepted interpretation in connection with elements of a reference. New knowledge evolves when new relations between existing knowledge elements are actively constructed by the human being, for instance, by introducing new relations in the reference context (also cf. the concept of the „linking metaphor" in Núñez, 2000). Subsequently, the old knowledge is brought into a logical consistency with these new knowledge elements – the hypostatized new relations (this disparity between new and old knowledge will be taken up in the following paragraphs and in section 4.2).

In a specific way, this description of what should be understood as really new mathematical knowledge will be referred to, in the empirical analyses, as a *criterion* in evaluating processes of interactive construction of mathematical knowledge. It will be used to decide, whether what has become apparent in an interaction was a successful new construction, or, on the contrary, merely a kind of regeneration of old knowledge and thus a rather unsuccessful knowledge construction. The qualitative, epistemologically oriented, empirical analyses of selected classroom episodes will contribute to the precision of defining this fundamental difference between old and new knowledge. At the same time, this criterion will help to clarify conditions of interactive constructions of mathematical knowledge. A practical application of this criterion of distinction in concrete mathematical interaction processes requires a clarification of how old and new mathematical knowledge can be identified in elementary teaching. What specific kind of knowledge plays the role of old or of new knowledge, and what are young students‘ personal descriptions of old knowledge and constructions of new knowledge?

Section 2.1 discusses the epistemological problem of the reciprocal relation between the socio-interactive and the epistemological dimension of the construction of mathematical knowledge. Starting from the paradox of old and new knowledge, that is from the epistemological problem of explaining how really new knowledge can be created out of old knowledge as a „logical combination“, a distinction between the „mathematical world“ in subject-independent, logical structures and the ideal mathematical objects (concepts) constructed by the human being, is introduced. Mathematical concepts are, according to this idea, social constructs in the sense that they are actively constructed in the culture of mathematics teaching. On the other hand, mathematical concepts are also assumed to possess a relatively autonomous (immaterial) existence, e.g. as „mathematical objects“ (Bereiter, 1994; Dörfler, 1995) which are embodied in the subject-independent „logical structure“ of the mathematical world. A central consequence of this theoretical conception is that the learning process of mathematical knowledge cannot be reduced to an individual and interactive adaptation to the social and cultural conditions of the class and the teacher. Learning, understanding and reasoning also has to take mathematical knowledge seriously, as a self-developing system that is autonomous to some extent. This theoretical interpretation provides, at the same time, an explanation for the „Learning Paradox“ with regard to the epistemological dimension.

Section 2.2 analyzes the relation between „social communication“ and „mathematical classroom interaction“. On the basis of the functioning mechanisms of general communication processes the characteristics specific to mathematical interactions in the classroom are identified. This particular communication contains a chosen object, the *mathematical knowledge* and besides it is a matter of a special form of communication, (intentional) *classroom conversations*. The semiotic interpretation of mathematical knowledge is examined in this theoretical approach using the epistemological triangle. The classroom interaction is influenced by the teacher‘s intentions, and this can lead to superimposing of the communication in which the semiotic interpretation intended by the teacher is reconstructed in communicative interaction patterns. Thus the two central dimensions of the research problem are put in relation to each other: the reciprocal relations between epistemological and socio-interactive conditions of the construction of new mathematical knowledge.

In section 2.3 the theoretical research question of the project is developed on the basis of the two dimensions: relations between subjective and objective conditions of mathematical knowledge (*epistemological* dimension) and the relations between general communications and special, mathematical classroom interactions (*socio-interactive* dimension).

The project examines the possible interrelations between epistemological conditions of the knowledge structure and socio-interactive constructions of meaning with the help of case studies, i.e. by means of qualitative analyses of episodes of everyday elementary mathematics teaching. The problem of examining interactively constructed forms of understanding and justifying general, mathematical statements with the help of symbol systems and reference contexts (in the context of the epistemological triangle) leads to the actual research problem. Mathematical justification of a general mathematical knowledge construct cannot be carried out by means of direct, finished assignments between „signs / symbols“ and „objects / reference contexts", e.g. by mere combination of existing knowledge representations. In order to interactively develop a generalizing justification, new relational connections – socially created mathematical objects (concepts) – have to be constructed and put into relation to each other again and again, in the structured reference context as well as in an appropriate symbol system, for being able to satisfy the intended generality of the statement.

In the interactive process of constructing a generalizing justification for a general mathematical statement by the creation of relations between „signs / symbols" and „objects / reference contexts" epistemological and social communicative conditions play an important role, and they are analyzed particularly with regard to the demand of the intended generalization:

- interactive negotiation of interpretations of elements of knowledge embedded in examples;
- agreed upon social conventions and epistemological necessity of the knowledge,
- use of the structured diagrams and arithmetical contexts for the generalizing justification.

The new, general frame for the construction of mathematical knowledge is created interactively, it is described and communicated and subordinate aspects of this knowledge are negotiated in the interaction, refused, accepted as useful etc. By means of the epistemologically-oriented interaction analysis of different examples of interactive knowledge constructions and forms of justifying, conditions are identified which play a central role in the possible success or failure of intended interactive processes of justification and generalization.

1. THE EPISTEMOLOGICAL PROBLEM OF OLD AND NEW MATHEMATICAL KNOWLEDGE IN THE SOCIAL CLASSROOM INTERACTION

Mathematical knowledge is facing a dilemma in the relation between old and new knowledge. On the one hand any piece of mathematical knowledge is logically consistent and hierarchically organized; thus new knowledge is logically deducible from the given foundations and thus, in principle, it is not really new on the basis of the logical structure. On the other hand there are really new and so far unknown insights in mathematics, among others, the solutions of previous unsolved problems and proofs of open conjectures. For example, the mathematical statement that there exist infinitely many prime numbers is a new insight for the human being, and, at the same time, it is a logical conclusion from the axioms of number theory and thus it is not really unknown on the basis of the hierarchical knowledge.

The example of mathematical proof best illustrates this paradox of mathematical knowledge. Rotman describes it in the following way:

> ... a proof is a logically correct series of implications that the Mathematician is persuaded to accept ... Such a characterization of proof is correct but inadequate. Proofs are arguments and, as Peirce forcefully pointed out, every argument has an underlying idea – what he called a leading principle which converts what would otherwise be merely an unexceptionable sequence of logical moves into an instrument of conviction. ... The leading principle corresponds to a familiar phenomenon within mathematics. Presented with a new proof or argument, the first question the mathematician ... is likely to raise concerns ‚motivation': he will in his attempt to understand the argument

> – that is, follow and be convinced by it – seek the *idea behind the proof*. He will ask for the story that is being told, the narrative through which the thought experiment or argument is organized. It is perfectly possible to follow a proof, in the more restricted, purely formal sense of giving assent to each logical step, without such an idea. ... Nonetheless a leading principle is always present – acknowledged or not – and attempts to read proofs in the absence of their underlying narratives are unlikely to result in the experience of felt necessity, persuasion, and conviction that proofs are intended to produce, and without which they fail to *be* proofs (Rotman, 1988, p. 14/15).

In an extensive historical and philosophical work, Jahnke analyzes the contradiction between knowledge development and knowledge justification and deduces didactic consequences for the learning of mathematics.

> In ... the way science sees itself ... »logic« and »intuition« are completely separated from each other. According to this conception new knowledge is gained intuitively and *then* codified and secured with the rules of logic. ... The process of gaining knowledge is thus of an essentially irrational nature or in the best case explainable as a psychological phenomenon, while, by contrast, the mathematical proof is only conceived as a tautological chain of signs (Jahnke, 1978, p. 58/59).

The analysis of this paradox between justification and development implies that it is only in the knowledge development that new „ideal mathematical objects" are constructed which allow for the change of the tautological, formal proof on the basis of the existing knowledge into a content-related justification „by looking back from the future". Thus, for example, the development of negative numbers throughout the history of mathematics is not a mere extension of the positive or natural numbers; on the contrary, the old, familiar positive numbers can now be interpreted from the point of view of the new conceptual aspects of negative or relational numbers in a generalized way „from the future". With reference to the problem of the „Idea behind the proof" or the proof necessity Jahnke says:

> ...the ... problem of conveying the need of justification of a theorem to children points to the fact that this can only happen under anticipation of more general aspects from which the possibility of an operative approach to mathematical circumstances or *idealized objects* becomes understandable (Jahnke, 1978, p. 251).

The core of this dilemma becomes clear with the following distinction. Mathematical knowledge can be conceived in two complementary ways. On the one hand each domain of mathematical knowledge can be understood as a logically consistent structure, in which all elements of the knowledge stand „equivalently" in logical connections to each other. On the other hand, in each such knowledge structure, one can identify or construct elements and concepts, which open up the way to new questions and problems that have not been logically embedded into the present structure yet (e.g. have not yet been „solved"). These elements thus make possible new insights.

This distinction between the *logical structure* and the *mathematical objects* corresponds to the philosophical distinction between a subjective ontology of reality and the subject-independent structures of the world established in this way (without being able to decide the problem of the „actual existence of reality" (Luhmann, 1997a, p. 218f.)).

> Ontology is the set of objects or actions in terms of which we experience the world or act upon it: tables, chairs, water, trees, cows, front, back, walking, swimming etc. Structure is the interdependence among the objects and actions, as in the cow being in front

of the tree. With this distinction in mind, I argue that while the ontology of the world is created by the cognitive agent, the structure of the world depends on the mind-independent external reality. In this way, the experiential world can be seen as both created and mind-independent at once. As there cannot be a structure without an ontology, it is the cognitive agent's act of creating an ontology that endows external reality with a structure. Moreover, the cognitive agent can change the ontology, thereby affecting the experiential structure of external reality. However, for any given ontology, the experiential structure of reality does not depend on the cognitive agent at all, and is determined by the mind-independent reality (Indurkhya, 1994, p. 106).

For mathematics, this means for instance the following: At the beginning of their development the first natural numbers are constructed by the human being with reference to real, different objects (e.g. pebbles): 1, 2, 3, 4, 5,.... The logical structure between all the numbers which determines the number system, e.g. in this case, the rule „there is always one more", cannot be arbitrarily changed by the human being. In order to illustrate the particularity of „logical structure" and „mathematical ontology" for mathematical knowledge, I consider the „mathematical world" of natural numbers. In this world of numbers, in the beginning as a world of quantities of different things, the human being can construct new mathematical objects, and thus create a new ontology of the mathematical world. But the human being cannot arbitrarily change the logical structures and relations. Such new objects are the mathematical concepts of „even number", „odd number", „square number", „prime number", „fraction", „decimal", „negative number", $\sqrt{2}$, and many other mathematical objects or concepts (for the possibility of different „mathematical ontologies" related to different fundamental standpoints, see Goodmann (1980); for a criticism on an individual-subjectivist interpretation of mathematical concepts, see Bereiter, 1994).

The subjective side of the creation of the ontology of the „mathematical world", during which the human being designates new mathematical objects, contains conventional aspects.

What counts as a correct application of the term ‚cat' or ‚kilogram' or ‚canyon' (or ‚klurg') is up to us to decide and is to that extent arbitrary. But once we have fixed the meaning of such terms in our vocabulary by arbitrary definitions, it is no longer a matter on any kind of relativism or arbitrariness whether representation-independent features of the world satisfy those definitions, because the features of the world that satisfy or fail to satisfy the definitions exist independently of those or any other definitions. We arbitrarily define the word ‚cat' in such and such a way; and only relative to such and such definitions we can say ‚That's a cat'. But once we have made the definitions and once we have applied the concepts relative to the system of definitions, whether or not something satisfies our definition is no longer arbitrary or relative (Searle, 1995, p. 166).

An important characteristic of mathematics is that, for example in the „world of numbers", the new objects are not immediately and directly perceivable. A prime number differs from a cat in that one cannot see the „defining property" of prime numbers (2, 3, 5, 7...) from the outside as an attribute, while the attribute of being a cat is visible. Prime numbers – like all new mathematical objects – are based on the *designation of a mathematical relation*, and not on the designation of a (visible) quality. For prime numbers this relation is e.g. „a natural number with exactly two divisors (1 and itself)". This special status of constructing new mathematical objects by „defining relations" becomes obvious from the fact that, in most cases, it can be extremely difficult to decide whether a given number is a prime number or not; this quality cannot be seen directly in the object.

Furthermore this particularity of the creation of mathematical objects (concepts) by the identification of relations contains the difficulty of raising only the „important“ and „central“ relations to the status of new concepts. Why do „prime numbers“ become mathematical objects and thus form a new ontology of the world of numbers? Why other numbers, defined by other relations, are perhaps not so interesting? In principle, any mathematical relation could be used to construct a new mathematical concept. Sometimes such a concept serves new insights for a certain period of time, until it is absorbed within a more general, extensive relation and becomes a part of a more abstract concept. Thus it becomes clear that complete conventional arbitrariness seems possible only in the choice of the name (prime number, even number, ...). The selection of relevant, central relations for the definition of mathematical objects and for the creation of a more developed ontology of the mathematical world is not purely arbitrary. It is, partly, subject to the conditions of the development of the insight-creating generalization. The importance of the new mathematical concept „prime numbers“ results from the idea that prime numbers constitute, in some respect, the elementary building blocks for all natural numbers within a multiplicative structure. The fundamental theorem of arithmetic states that every positive integer greater than 1 can be written uniquely as a product of prime numbers (with the prime factors in the product written in non-decreasing order) (cf. Eccles, 1997).

The distinction between the *subject-independent logical* structure and the *ontology created by the human being* of the „mathematical world“ has direct implications for explaining, understanding and justifying mathematical statements in social interaction. The dilemma between old and new knowledge represented above contains a new problem in the frame of mathematical interactions. This will be illustrated now by means of a typical case of interaction.

In a third grade class students learn addition and subtraction strategies in the number space up to 100. Besides the standard arithmetical operations, in certain exercises students are expected to use conceptually-oriented strategies with important connections specific for the given numbers (cf. Wittmann & Müller, 1990, p. 82ff). For addition and subtraction, such strategies could be based, for example, on the principles of *constancy of the sum* or *constancy of the difference*. The application of such strategies could look, for instance, as in the following calculations:

$$23 + 58 = 21 + 60 = 81 \text{ or: } 20 + 61 = 81$$

$$\text{and: } 63 - 19 = 64 - 20 = 44$$

The fundamental idea here is that the two terms of the sum or the minuend and the subtrahend are changed in such a way that one of the two becomes a multiple of ten (the units position is zero) and thus the result of calculation is obtained efficiently and quickly.

The „world of numbers“ observed here consists, among other things, of the correct „logical structure“ to further develop addition and subtraction up to 100. One can say that with the relations of the constancy of sum and difference new „mathematical objects“ (local concepts) are introduced and meaningfully conceived. In this way the ontology of the number world which was created by the learner can be enriched.

The interaction between the teacher and a student occurs in the following steps. The teacher explains this strategy and tries to illustrate its core idea to the student. Since the student does not understand the conceptual relation, the teacher goes back to an easier strategy by inserting an intermediate step. She explains the rule to the student according to the following calculation steps:

$$23 + 58 = 23 + 60 - 2 = 23 - 2 + 60 = 21 + 60 \text{ (alternatively: } = 83 - 2 = 81\text{)},$$

and the other rule:

$$63 - 19 = 63 - 20 + 1 = 63 + 1 - 20 = 64 - 20 \text{ (alternatively: } = 43 + 1 = 44\text{)}.$$

In the dialogue with the student the teacher draws her attention to the fact that in the first exercise one has added a little too much and thus has to subtract just as much. And in the second exercise one has subtracted a little too much and thus has to add just as much. The student, however, does not understand the *actual* meaning of this relation which conceptually consists of its regularity as a constant result: One transforms the task into another task, which is arithmetically equivalent but easier to calculate. The student orients herself by the step-by-step working out of the rule and thus the mathematically correct procedure.

The request to give reasons why something has to be subtracted or added is finally answered by the student in the following way: „Yes, with plus I have to calculate minus and with minus I have to calculate plus." She formulates a mnemonic rule for the procedure without being able to deal with the conceptual relation behind this procedure. The teacher is aware that the student is trying to evade the answer. She therefore continues her attempts to make the arithmetical meaning of this calculation strategy accessible to the student. Despite all her efforts, she fails in achieving her goal.

A question by the teacher makes the student feel insecure and she expresses her doubts by asking: „No, or do I, rather have to calculate minus with minus and plus with plus!??" She lacks the conceptual understanding to justify the strategy. The more questions the teacher asks – and she has to ask them since the student obviously has problems understanding the explanation of the strategy – the more the student is overwhelmed with doubts, which in turn causes the teacher to go back to isolated numbers and to operate on them in little steps. Thus a possible coherent view on the conceptual relations and patterns between the numbers dealt with is more less likely.

In the course of the interaction the teacher increasingly tries to explain the intended conceptual calculation strategy by the logical structure of the correct additions with further intermediate steps. Thus their actual meaning is taken from the new concept relation „constancy of the sum or of the difference". This new mathematical object is explained by a reduction to the familiar, logical structure of correct addition steps. This way the logical correctness of this mathematical relation is „proved". But, through this one-sidedness, the new object is, at the same time, destroyed.

This example leads to the following observations. There are two kinds of understanding or apprehending, and, accordingly, of explaining and giving reasons for mathematical statements. On the level of the subject-independent, consistent structure, the construction of a logical, mathematical connection between the old and the new statement leads to a formal equality and thus correctness, which then induces a formal, *understanding* or better *apprehending of the logical connections*. The learner can largely „reproduce“ this logical insight instead of subjectively interpreting it. One understands, explains or gives reasons for a mathematical statement by reducing its logical correctness to the subject-independent structure. In this case, however, one often achieves not an actual understanding, but a formal equality of statements.

On the other hand, there is *ontological understanding*. The new relation is constructed as a new mathematical object by the individual themselves. This relation is not just reduced to its correctness in the frame of the logical structure, but is holistically represented as a new relation. This procedure of the construction of mathematical objects or concepts by the identification of mathematical relations requires manifold experiences, experiments and attempts of children in the world of numbers. Furthermore the young students must have confidence that a personal creation of mathematical objects is possible, even if one often has to carry out changes and new constructions until one obtains a „direct“, subjective, ontological insight into new mathematical content.

Both forms – *logical* and *ontological* understanding and reasoning – are complementary and they need each other. Logical understanding or apprehension of the logical connections alone does not allow the human being to gain personal insight. The person is limited to the requirement of a formal reproduction of the logical structure conditions. The ontological, subjective construction of new mathematical objects (concepts) remains unfinished as long as these new objects are not correctly connected in a logical structure.

The social interaction of mathematics instruction deals differently with *logical* and *ontological* understanding and giving reasons. Logical connections and giving reasons (logical correctness) can be communicated to the other participants in the interaction step by step. Ontological issues, explaining and giving reasons, on the other hand, cannot be directly accessible or visible to the other participants. Ontological understanding and giving reasons always require an active interpretation by the learner which can lead to the construction of new mathematical objects on the foundation of a common interaction which then bestow the „mathematical world“ with a differentiated, subjective ontology and in this way can increase the understanding of the individual.

The question, for instance, of whether the number 31 is a prime number or not, can be examined step by step and communicated by the defining relation „a number which can only be divided by 1 and itself“: One divides the number by 2, 3, 5, 7, and finds out that there is always a remainder, thus the number 31 cannot be divided by numbers other than 1 and 31. The subjective ontology of the object (concept) „prime number“, on the other hand, is not reducible to the defining quality „a number which can only be divided by 1 and itself“. Fundamentally, the ideal mathematical object intended by this defining quality is inexhaustible and cannot be fabricated as a finished, completed object. A hint towards this inexhaustibility of the object „prime number“ are the innumerable arithmetical relations, statements and qualities which can be related to this concept.

Making possible a subjective ontological understanding, based on the interactive construction of mathematical objects, is a core problem of mathematical teaching and learning processes. When understanding and giving reasons in the classroom is limited to logical understanding, because only this can be communicated in the interaction, there is the danger of obstructing the ontological understanding in this way or of making it even completely impossible. All teachers are aware that new numbers and new concepts have to be made „understandable" to children. Yet in the explanatory processes teachers especially emphasize the subject-independent logical structure of the „mathematical world" – the only thing that can be directly communicated in the interaction – and they underestimate the necessity of the construction of a subjective ontology which can only be actively created by the learner themselves.

2. THE RELATION BETWEEN „CLASSROOM COMMUNICATION" AND „MATHEMATICAL INTERACTION"

The realization and the progress of communicative systems is an insecure or uncertain event at the beginning. According to Luhmann, communication consists „... in the synthesis of three selections: (1) message; (2) information; (3) understanding of the difference between message and information" (Baraldi, Corsi & Esposito, 1997, p. 89). These elements are possible reasons for the uncertainty of the event communication.

> Communication is ... an improbable event. It possesses three levels of improbability. On a first level it is unlikely that the communication is understood – i.e. that it occurs at all. On a second level, more rich in prerequisites, it is unlikely that the message reaches the addressee. In even more complex situations it is eventually unlikely that the communication is accepted (received) (Baraldi, Corsi & Esposito, 1997, p. 93).

In order to explain the realization of communication in different contexts such as for instance science and education, Luhmann develops the concept of the „symbolically generalized communication media".

> In general, symbolically generalized communication media are semantic institutions which make the success of improbable communications possible. ‚Making the success possible' means in this context: increasing the readiness for acceptance of communication in such a way that communication is risked rather than given up as hopeless from the very beginning (Luhmann, 1982, p. 21).

According to Luhmann symbolically generalized communication media are not concrete institutions but abstract semantic directives and rules of interpretation.

They are general and do not depend on concrete situations, i.e. they can be used to communicate in a variety of situations.

> An example for this is science or the communication medium belonging to it, i.e. »truth«. Scientific communication resorts to the communication medium of truth; it thus signals that its messages are the result of scientific procedures and not only of opinion or appraisal. This blocks counter arguments and thereby increases the acceptance of statements, even if they are contrary to intuition and experience (Heintz, 2000, p. 250).

Against this background, I will explain in the following, the functions of communication in mathematical science and which role takes the concept of the generalized communication medium. This description will then serve to illustrate the particular characteristics of communication in the school institution and especially to contribute to a better understanding of the relation between „classroom communication" and „mathematical interaction".

The first communication problem for mathematics consists in the fact that the mathematical objects or concepts by means of which mathematicians communicate with each other are not items that can be perceived by the senses (cf. section 1.1, Duval, 2000; Heintz, 2000, p. 219). This problem corresponds to the epistemological dimension of mathematical knowledge as it has been developed in section 1.1. In mathematical science the communication problem also becomes elevant as a relation between the „psychological" and the „social" (cf. section 1.3).

> It concerns the relation between thinking and communicating, between ‚consciousness' and ‚communication'. ... The complaint that there is an unbridgeable gap between the personal world of thoughts and what can be communicated can often be heard in mathematics. ... The fact that this mediation problem is especially important for mathematics also depends on the way of working which dominates there. Mathematicians often treat a problem by themselves alone and for a very long period of time. ... The gap between the private world of thoughts and the public communication can, circumstances permitting, become so deep that communication of thoughts is regarded as nearly impossible (Heintz, 2000, p. 220/221).

This second communication problem of mathematical science generally corresponds to the problem of communication as an autopoietic system in reciprocal relation with its necessay environment, the psychic system (cf. section 1.3).

In view of these two communication problems, how is it still possible that in mathematical science „success is made possible" for communication between mathematicians and that mathematical communication occurs? In general the symbolically generalized communication medium for science consists in „truth"; so truth is at first also the medium which makes mathematical communications highly likely. It has to be further explained how truth is established in mathematics. In her extensive work Bettina Heintz (Heintz, 2000) argues that it is the mathematical proof which plays the role of the symbolically generalized communication medium for mathematical science and solves the two above mentioned communication problems.

> With the institution of proof, mathematics has constructed a highly elaborated framework of norms with the purpose of making mathematical communication easier (Heintz, 2000, p. 220).

One the one hand, proof helps in communicating about the invisible mathematical objects. Furthermore, proof mediates between consciousness and communication. Writing down a proof is a way of translating thoughts into communicable statements.

> This is exactly what the proof does. The translation itself usually occurs while writing it down. ... In the frame of this work of translation the proof has an important task of regulating the behavior of speaking. The proof is, in a certain way, the hinge which mediates between consciousness and communication, between the ‚psychic system' and the ‚social system'. No other discipline has set up as detailed communication rules as mathematics with its institution of proof (Heintz, 2000, p. 221).

In its long historical development the modern mathematical proof has become the symbolically generalized communication medium specific for mathematics which makes mathematical communication possible and successful. Especially the transformation of proofs, which are founded on common and intuitive pre-knowledge to formalized proof procedures, is an essential condition.

> Even when arguments are deductively constructed and argue with the help of logical rules, when these arguments require a common knowledge and rely on intuition and visualization they are more at risk of dissent than a formal argumentation which one can hardly avoid even if it is contrary to intuition and experience. Exactly this mechanism is described by Luhmann with the concept of the symbolically generalized communication medium. A shift to communication media occurs when socio-integrative mechanisms – direct communication and common normative background – are no more sufficient to create confidence and to guarantee the connectivity of scientific communication. This argumentation suggests regarding formalization and informal interaction as historical alternatives up to a certain degree. The formalized proof only becomes an important communication medium when the informal mediation of mathematical knowledge fails (Heintz, 2000, p. 274).

How does Luhmann's theory explain the possibility of the realization and progress of communication processes in the frame of the education and teaching system? For educational intentions – first of all to be distinguished from subject matter oriented instruction – the following applies.

> The educational system is a partial system of modern society ... with the function of ... starting changes in the single psychic systems, so that these can also participate in more unlikely communication which society (re)produces and which occurs mainly in the other function systems. The particularity of the education system thus consists in the fact that its function is not primarily aimed at the processing of communication or the generation of communicative consensus, but at the change of the psychic environment of the society. The effects of the education appear outside of the society, in the abilities and skills of the individuals, that means in their competence of participating in the communication (Baraldi, Corsi & Esposito, 1997, p. 50)

Since communication in the frame of education systems differs from communication in other social systems, there are no corresponding symbolically generalized communication media.

> That is why there is no ... symbolically generalized communication medium which makes the success of educational communication more likely because not even such media can operate in the environment of society: There is no chance of motivating the single individuals who are to be educated to accept the teacher's educational intention and to orientate their behavior according to the teacher's expectations (Baraldi, Corsi & Esposito, 1997, p. 50).

But still, subject matter oriented instruction works in a comparable way, so that the communicative exchange between teacher and students is initiated and continued over a certain period of time. The education happens in institutionally organized instruction.

> A further particular characteristic of education consists in the fact that it only functions if interactions between teachers and students in the classrooms can be organized regularly. School interaction is a functional equivalent of the here absent symbolically generalized communication medium, because it creates situations in which socialization ... is forced in a very improbable way, and this improbability permits education to plan and if necessary set off well-aimed effects in the students' consciousness systems (Baraldi, Corsi & Esposito, 1997, p. 50).

Everyday mathematics teaching is thus placed in the following tension. As subject matter oriented instruction which relates to mathematical knowledge, the success of the instructional communication is at least partly tied to the criterion of the mathematical truth or the correctness of mathematical statements. On the one hand, the emergence and progress of mathematical classroom communication – as long as it is possible under educational conditions with its particularities – is regulated and forced by the interactively negotiated correctness of mathematical statements. In contrast to scientific mathematical communication, however, in school-mathematical interaction one has to take into account the fact that students are not as familiar as mathematical experts with the communicative rules of the formal proof, and that they cannot use these rules as a mathematical communication medium in the same way (cf. Heintz, 2000). Instructional mathematical interaction is expected to contribute to introducing individuals into this mathematical communication practice, and thus to increase these individuals‘ ability to participate in (mathematical) communication in the society (see the above-mentioned quotation of Baraldi, Corsi & Esposito, 1997).

One the other hand, one has to be aware that mathematical interactions in the classroom between teacher and students are largely argumentations which – in Heintz‘s description – „suggest a common knowledge and appeal to intuition and visualization“ (Heintz, 2000, p. 274). The decision over the possible correctness of statements would thus be subject to certain ambiguities and controversies, but, in the frame of the classroom interaction, these are usually decided by the teacher. This evaluation carried out by the teacher in terms of what is mathematically right and what is mathematically wrong is an essential selection criterion of instruction.

> The teacher never ... knows which effects his pedagogical behavior can have; he can only observe how the students behave, and evaluate the difference or non-difference from his expectations. In this sense education has the possibility of selection, that means the production of evaluations because of the difference between improvement and deterioration of the students‘ achievements (Baraldi, Corsi & Esposito, 1997, p. 50).

Mathematical interactions are thus social systems which are characterized, on the one hand, by particular intentions (e.g. of education) and, on the other, by their subject matter, which is some form of mathematical knowledge.

Classroom activities between teachers and students are socially and pedagogically intended interactions with the aim of mediating knowledge. This implies a superimposition of the autopoietic development of the social classroom communication by an „additional meaning“, namely by the teacher‘s implicitly dominant effort to make the students‘ success in learning more probable by means of interactive measures. This effort is perceivable (and evident) for all the participants in the communication for both teacher and students. When trying to realize their instructional intentions in the mediation of knowledge, teachers often unconsciously proceed on the assumption that the gap between the social and the psychological system, between communication and consciousness (between thinking and communicating, Heintz, 2000, p. 221) could be almost directly closed, and that the communicated meaning could be directly transported into the students‘ consciousness. A familiar example for such an instructional course of interaction is the so-called funnel pattern (cf. among others Bauersfeld, 1978; Krummheuer & Voigt, 1991; Wood, 1994; 1998).

Krummheuer and Voigt describe this pattern in the following way:

> The teacher starts with an open question. Because of the diverging answers by the students the teacher feels obliged to make his questions more and more narrow until the classroom conversation is canalized to the difference of the results and finally to the saving key-word (Krummheuer & Voigt, 1991, p. 18).

The following short classroom communication represents a typical example of the final, strictly confined, phase of a communicative funnel. In a second grade class, students have to calculate the exercise: „40 + 16 = ?“

1	T.:	Equals? Sebastian?
2	Sebastian:	Equals, ehm, 66.
3	T.:	So, 6 ones and how many tens?
4	Sebastian:	66.
5	T.:	But why? How many tens do you have here?
6	Sebastian:	4.
7	T.:	And here?
8	Sebastian:	6, 16, …
9	T.:	How many tens?
10	Sebastian:	One.
11	T.:	One and here 4 tens, and how many tens do you then have together?
12	Sebastian:	5.
13	T.:	5, and the 6 ones here, here are no ones.
14	*Teacher writes the result „56“ on the black board.*	

It is only by forcing the correct numbers with the help of key-words such as „tens“ and „ones“ and by posing supporting questions that the teacher obtains the result. In Bauersfeld‘s words: „Further absence of the expected answer leads to the narrowing of the teacher‘s efforts to the mere reciting of the expected answer …“ (Bauersfeld, 1978, p. 162). And further: „Open to some extent in its beginning, the pattern becomes stabilized from turn to turn, the freedom of choice dwindles, and there is restrictive power with the process in the end“ (Bauersfeld, 1988, p. 37). The funnel is „a narrowing of the actions by expectation of an answer“.

The interaction between the students and the teacher often works so well that students no longer relate the teacher‘s statements (her signifiers) to the mathematical aspects of the problem in question, but interpret them totally in the context of the interactive hints, allusions and expectations. This interactive game works so well because children can meet the teacher‘s expectations quite easily, and because the teacher seems to succeed in satisfying her instructional intention of effectively imparting knowledge to the children. These and comparable interaction mechanisms between teacher and students can actually take the role of the communicative rules in the classroom which makes the realization and the continuation of the classroom communication possible.

Mathematics teaching is, however, concerned with mathematical knowledge, and thus the classroom processes are also influenced by aspects of the mathematical correctness. For someone who observes the mathematical classroom interaction from the outside, e.g. the researcher, it is necessary to analyze the knowledge, commonly constructed in the interaction process and also from an epistemological perspective, as has been argued in the previous chapter. The analysis of the special status of school mathematics (see section 1.1) and its interactively constituted meanings gives rise to not merely seeing the classroom interaction as a functional equivalent of the symbolically generalized communication medium (Baraldi, Corsi & Esposito, 1997, p. 50). In order to examine and understand the way mathematical classroom communications can be initiated and continued, one has to also consider the school-mathematical correctness, including the communicative rules according to which the correctness in the mathematical interaction is created. Accordingly, in the following, I shall present „school mathematical correctness", by analogy to scientific truth (Baraldi, Corsi & Esposito, 1997, p. 190) and the formalized proof (Heintz, 2000, p. 274), as a triggering factor for the realization and continuation of mathematical classroom communication. Thus „school mathematical correctness" will be understood as a functional equivalence of a symbolically generalized communication medium in school interaction.

The comparison with scientific mathematical communication can even be further expanded. In order to realize and keep classroom communication going one needs a balance in the fundamental tension between a general classroom communication based on interactive mechanisms and patterns and a (school-) mathematical way of argumentation about correctness and falseness of school-mathematical statements. A form of symbolically generalized communication medium analogous to mathematical research which could guarantee the progress of mathematical classroom communication, as is the case for the formal proof in mathematics, cannot exist in the context of instruction. The reasons for this are twofold. On the one hand, school interaction is the (didactic) means by which students are introduced into the rules of mathematical communication. This, on the other hand, requires that school interaction uses not only internal strategies just to keep the communication going, but that connections are constantly made to mathematical communication rules. This suggests conceiving of mathematical classroom communication in the ideal case as a process which consists of interrelationships between the necessary school-specific interactions – as an educational form of introduction in not yet practiced mathematical communication forms, – and suitable school-mathematical argumentation forms to decide between right and wrong – as the particular subject matter oriented communication content with its particular epistemological qualities.

Qualitative analyses of everyday mathematics instruction show that this balance is very difficult to obtain. Many prerequisites have to be satisfied. Therefore, everyday mathematical classroom communication is often one-sided. „The mathematical logic of an ideal teaching-learning process ... becomes replaced by the social logic of this type" (Bauersfeld, 1988, p. 38). The contrast between direct pedagogical intentions and epistemological conditions of emergence of mathematical knowledge in classroom communication can lead to the following *paradox of instructional mathematical communication:*

(Mathematical) information is to be transmitted by the teacher's message, which is not directly achievable and which is, in principle, completely impossible. This is why *communication about the content of mathematical knowledge* is often „transformed“ into *communication about the information intended by the teacher*.

These „communicative substitute strategies“ could be observed in episodes analyzed in this research project. In these cases students would not interpret the (mathematical) signifiers / signifieds used in the communication with reference to other *mathematical* signifieds, and thus would not create (elementary) *epistemological* relations in the interaction, but they would try to decode the interpretations attributed to the teacher in order to be thus able to create the supposedly correct relation. In mathematical interaction one can therefore observe that the possibility of the autopoiesis of the communicative, mathematical system emphasized by Luhmann is being „circumvented“: The receiver of the message (a student) is at first only able to attribute the given message to the messenger (for instance the teacher).

Being able to detach the messenger‘s information from the message given by them means the precise possibility of the autopoiesis of the social system of mathematics teaching. (cf. section 1.3). But in teacher-students mathematical interactions one can observe how, in another – a „superimposed“ – social communication system, students in interaction with their teacher try to tie the teacher‘s information to his message, i.e. to deduce the teacher's information from his message. This communication occurs under the tacit assumption that the mathematical messages (the signifiers / signifieds) relate to *definite* information (fixed referential signifieds) which can be discovered in the communication *above* these messages. The messages of the teacher (and of other students) become the content of communication themselves and they are not given in the communication only in order to interpret these interactively with the help of other (mathematical) signifiers / signifieds.

In the analysis of mathematical social interactions it is thus necessary to distinguish between an epistemological system and a socio-interactive system, and examine the possible interrelationships between them.

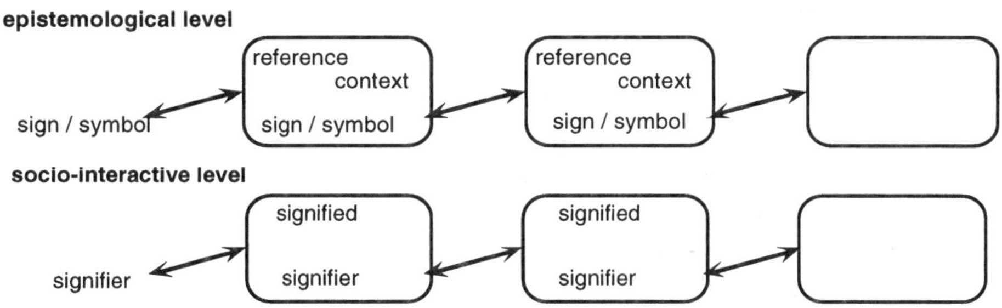

Figure 14. Correspondences between epistemological and socio-interactive level.

The socio-interactive plane of analysis is focused on how signifiers / signifieds can be „spontaneously“ interpreted by the participants by reference to other signifiers given in the communication, and how the autopoiesis of the communication is initiated. The epistemological plane of analysis is related to the questions of „truth“ or „school-mathematical correctness“. It raises the question of assignment of meaning to the mathematical signs / symbols, in order to explain why certain interpretations are accepted or rejected on the social plane. The two planes are interrelated: The interactions on the social plane „produce“ interactive, epistemological interpretations of the mathematical signs / symbols. The epistemological meaning assignments of the signifiers which can be observed and analyzed in the reflection of the social interaction are necessary to evaluate the reasons for the success or the failure of mathematical communication (for that which is constituted as true or not true in the mathematical communication, and which has to be distinguished from „immediate" mathematical truth or falseness).

In contrast with de Saussure's semiotic model with its terminological distinctions between „signifiers" and „signified" which constitute the „sign", a different notation has been chosen for the epistemological triangle: „sign / symbol", „object / reference context" and „concept" (see section 1.1). This terminology takes into account the fact that mathematical signifiers are often understood in the form of a sign themselves which represents an aspect of a reference context. Furthermore the mathematical concept is of a central epistemological meaning in mathematics exceeding the function of being a „sign / symbol". Once again I stress the conception that the mathematical signs / symbols should not be confused either with the mathematical concepts (ideal objects) or with the signified reference contexts, but that I proceed on the assumption that the three entities involved are mutually autonomous (see section 1.1). Hence, before developing the background to the communication of the instruction and to the epistemology of the subject mathematics, I assume a dyadic semiotic model on the one hand and the triadic epistemological triangle on the other.

The comparison and the partial correspondence between the semiotic model and the epistemological triangle aims at making it clear, among other things, that the general, communicative foundations of the epistemological analysis of mathematical interactions are provided by Luhmann's conception of communication as an autopoietic system. Moreover, the conceptual notations in the epistemological triangle are a way to do justice to the particular conditions of the emergence of mathematical knowledge. The perspective of the semiotic model aims at the general conditions of social communication in its dynamic development;. The perspective of the epistemological triangle reflects the particular status of mathematical knowledge as it has been constructed in the interaction to a certain point of time as a specific, social fact which is characterized especially by the symbolization of *relations*.
The epistemological triangle allows uniform characterization of the mathematical object (the concept) as a symbolized mathematical relation in the dialectic of socio-subjective ontology and socio-independent, logical structure. In a differentiated interpretation of the triangle according to which the referential connection between „sign / symbol" and „object / reference context" is not understood as the assignment of names to things, but as a reciprocal reference between relational networks, the aspects of the logical structure of the mathematical concept are emphasized by the „sign / symbol" vertex, and the conditions of the subjective ontology in the form of identified relations to the signs / symbols are emphasized by the „object / reference context" vertex.

3. SOCIAL CONSTRUCTION OF NEW MATHEMATICAL KNOWLEDGE STRUCTURES FROM AN INTERACTIVE AND EPISTEMOLOGICAL PERSPECTIVE

In the following chapter, I will use the theoretical perspectives on new mathematical knowledge, presented in the previous sections, to characterize and analyze different patterns and types of justifying and generalizing knowledge construction which occur in typical episodes of everyday mathematics teaching (in grades 3 and 4 of elementary school). (For a conception of the content of mathematical instruction and for the realization of the documentation of observed mathematical lessons see chapter 3.) In the analysis, I will be using two theoretical dimensions, one aligned with the *communicative interpretation* of the mathematical knowledge, and the other with the *epistemological characterization* of the interactively-created knowledge.

The interactive construction of new knowledge is a fragile process in the sense that, ultimately, its success cannot be forced or guaranteed. Like any creative, constructive act, the creation of new knowledge is subject to the continued effort of producing something as yet unknown and not yet existing in this form. However, one might think that in teaching, construction of new knowledge could be a methodically guided and successfully organized process rather than a free creative act. In fact, it is possible as well as necessary to support the process of teaching and learning of new mathematical knowledge in such a way that success becomes likely and does not remain completely arbitrary. This is attempted, for example, in the design of the so-called substantial learning environments (SLE) (Wittmann, 2001) whereby children are supposed to make their construction process of new knowledge more effective by means of their own activities (cf. section 3.1). In spite, however, of all methodical and didactic measures the instructional construction process of new knowledge remains, fundamentally, a fragile, indeterminate venture.

The construction of new knowledge in mathematics teaching occurs under two important conditions: The particular character of classroom communication („Communication with the teacher who knows superimposes mathematical interaction") and the particular epistemological nature of mathematical knowledge („Mathematical knowledge consists of symbolized, operational relations") (see sections 2.1 and 2.2). One consequence of these conditions is that in instructional processes the actual construction of new knowledge remains, ultimately, an open-ended joint interactive process whose outcome cannot be completely determined. If the teacher who knows tries to communicate new knowledge to the students in a more or less direct way, this very action devalues and destroys the new knowledge as new knowledge (this is a similar issue as the „Topaze effect" described by Brousseau, 1997, p. 25). New knowledge as a symbolic, operational relation to be identified cannot be communicated directly (section 2.2); the only thing that *can* be communicated directly is how a ready made mathematical object fits into a certain logical structure. Because of this particular epistemological character of mathematical knowledge, an explicit message (for example a mathematical definition) about some new mathematical knowledge cannot contain the open, multiple meanings or interpretations of the explicit mathematical signs and rules. The interpretations of the mathematical signs and rules (e.g. understood as signifiers for which the learner has to construct the signifieds in social interaction) have to be created subjectively and are not fully determined, but are manifold and, in principle, open for new interpretations.

Thus the construction of new mathematical knowledge in the processes of teaching and learning is a fundamentally interactive creative proceeding whose fragile and open outcome has no warranty of success. Successful constructions of new knowledge by students and the teacher are not the result of deterministic cause-effect processes, but rather spontaneous events (Wood, 1996, p. 103). The central point here is that with every new construction of mathematical knowledge one has to struggle for a new interpretation, by exploration, using paraphrases and explanations by means of comparisons. This uncertain, interactive search has to take into account the epistemological nature of mathematical knowledge (symbolized relations) and the conditions of the interaction (superimposed or open communication).

The communicative and epistemological conditions are used as two central theoretical dimensions of analysis of episodes of interactive construction of new mathematical knowledge. The first dimension, „communicative interpretation of the mathematical knowledge“, focuses on the tension between the (direct) conveyance of factual knowledge and the construction of interpretations of the new knowledge. The second dimension, „epistemological characterization of the mathematical knowledge“, refers to the tension between the (empirical) situatedness and the (relational) generality of the new knowledge.

In the interactive construction of mathematical knowledge, the communicative dimension appears in the tension between the directly communicable concrete qualities of mathematical knowledge (the logical, deductive representation of facts and rules) and the new interpretations of the mathematical content as symbolic, operational relations and conceptual aspects (section 2.1). The construction of really new *mathematical* knowledge requires to focus on a mathematical relation as „the“ new mathematical object which has to be interpreted, at first, as an ontological entity and then embedded in the consistent logical structures of the existing knowledge. To sum up this dimension can be represented as follows:

Communicative Interpretation of the Mathematical Knowledge	Mediating the pre-given properties of the object of Communication:facts, rules, logical connections	Balance between *Mediation of Facts* and *Construction of Interpretations*	Construction of potential interpretations of the object of communication: relations, symbols, conceptual aspects
Characteristics of the Communication	• logical verification, backtracking • “closed“ (direct) communication • teacher communicates completed knowledge		• new construction, • “open“ (indirect) communication • students try to describe the conventional aspects of the new knowledge

Figure 15. The communicative dimension.

In the mathematical communication between the teacher and the students, possible reasons for new knowledge and its construction can only be achieved if the interaction leaves open the participants' (especially the students') possibilities of interpreting and making connections, and if the teacher's intention (the „teacher's sign") is not *re*constructed by the students, but, instead, an autonomous, social signified for a signifier is actually interactively *constructed*.

The second dimension refers to the epistemological nature of mathematical knowledge. The central idea here is the tension between an initial empirical interpretation of elementary mathematical concepts and an understanding that mathematical concepts embody relations and structures in a symbolized and operational way. Any step of construction of new mathematical knowledge necessarily involves some mathematical generalization, however limited in scope. Otherwise, only facts, „quasi empirical" facts, are stated at the logical knowledge level (e.g., children give justifications such as, „the calculation or the result is right"; or the statement is correct for typical concrete cases). To go beyond the establishment of empirical facts, new, meaningful relations have to be constructed.

The relation which is the topic here is discussed in the literature as the contrast between the *situatedness* and the *universality* of mathematical knowledge (cf. Bereiter, 1997). According to the interpretation proposed here the problem of the fundamental context dependence of mathematical concepts and mathematical knowledge is unquestionable (cf. Steinbring, 1998a). The critical point in this relation is the type of context dependence: Is the epistemological interpretation of the mathematical knowledge immediately tied to the concrete, empirical qualities of the situation, or does the situation with its structure serve as an open reference context which has to be interpreted first and which allows for new interpretations? From this point of view, situatedness of mathematical knowledge is understood as a relation between signs / symbols and reference context, whereby the given context does not directly explain the knowledge, but the situated context can be re-interpreted and used as an embodiment of structural connections which allow for constructing new mathematical knowledge. To sum up, the epistemological dimension can be represented as follows:

Epistemological Characterization of the Mathematical Knowledge	Empirical, situated characterization of mathematical knowledge	Balance between *Situatedness* and *Universality*	Structural, relational universality of mathematical knowledge
Role of mathematical signs and symbols	names for empirical things and qualities		embodiment of mathematical relations as exemplary „variables"

Figure 16. The epistemological dimension.

The central aspect of this epistemological dimension is the fact that neither the (concrete, situated and structural) reference contexts nor the sign and symbol systems directly contain the mathematical knowledge or the mathematical concept. For instance, the number concept cannot be reduced to quantities of countable, concrete or abstract objects, nor to the numerals which are used for their encoding. The nature of the theoretical, mathematical object „number concept" does not consist in the concrete qualities and forms of the formal signs and corresponding reference contexts. Signs and reference contexts form a possible foundation for the construction of aspects of a mathematical concept in so far as they make it possible to hint at structural connections and can be interpreted as embodiments of these aspects. Mathematical signs and reference contexts do not directly and immediately depict the new knowledge. But they serve as necessary iconic carriers of the knowledge in the sense of indicators of other structural relations of the concept. When coping with the relationship between mathematical signs and corresponding reference contexts students in mathematics classes are faced with the specific problem of interpretation in such a way to always detach themselves from the concreteness of the situation. They are requested to see, interpret or discover „something else", another structure, in the situation. This problem is discussed in mathematics education literature under the recurring theme of „discovering the general in the particular" (cf. Cobb, 1986; Mason & Pimm, 1984).

From an epistemological perspective, with this problem, there is a strong emphasis on the fragility and openness in the construction process of mathematical knowledge (cf. Brousseau & Otte 1991). In the historical development of mathematics the concept of variable (in elementary algebra) has emerged as a prototype for the operative notation of the mathematical processing of the new, unknown and still open knowledge. Students in elementary school are obviously far from using the concept of variable in their construction of mathematical knowledge. In elementary mathematics idiosyncratic metaphorical notations, specific verbal references to generalities and general patterns in special given (geometrical or arithmetical) diagrams and contexts can be used for a description anticipating the new knowledge.

Processes of the construction of new mathematical knowledge are subject to an interplay between structure and object. On the one hand, a relevant relation has to be identified and anticipated. On the other, the mathematical relation is „hypostatized" by means of this process in order to be operationally anchored in the logical structure of the existing knowledge and also to be used productively. Furthermore there are problem contexts in which hypostatized relations have to be „resolved" back into their relational structure. Thus there occurs a flexible, varying change between a way of reading of mathematical entities as „quasi empirical" operational, mathematical objects and as relational, structural constructions (cf. Cassirer, 1957). The consequence is that the development of mathematical knowledge does not represent a permanent and irreversible abstraction on increasingly higher levels, but, fundamentally reflects the reciprocal relation between situatedness and universality of the mathematical knowledge. Lakoff and Núñez described the relation between the effect of the development of the mathematical knowledge and action on the objective world. The original actions and experiences of the human being in the objective world represent the „fundamental metaphors" of all human acting and thinking out of which scientific concepts – thus also mathematical concepts – develop and are constituted (Lakoff & Núñez, 1997). Hence one cannot fundamentally escape from the embodiment of theoretical concepts. The mathematical thinking of a human being must permanently expose the „quasi empirical" operative units, which were hypostatized in a process of generalization, to new, potential relational structures and thus give them new, meaningful interpretations.

The classification of selected teaching episodes, in which interactive construction processes of mathematical knowledge were taking place, will be carried out using the following two-dimensional matrix (Fig. 17). The interplay of socio-communicative and epistemological aspects in the construction of mathematical knowledge will be used in this model as the most important level of interpretation. The application of this evaluative model is based on the three steps of analysis elaborated above:

- description of the episode along the single phases
- development of the epistemological interpretation in the course of the interaction
- detailed analysis of reasoning patterns from the communicative and the epistemological points of view

Communication / **Epistemology**	Mediating Pre-Given Properties of the Object of Communication: Facts, Rules, logical Connections	Balance between *Mediation of Facts* and *Construction of Interpretations*	Construction of Potential Interpretations of the Object of Communication: Relations, Symbols, Conceptual Aspects
Empirical, Situated Characterization of Mathematical Knowledge			
Balance between *Situatedness* and *Universality*			
Structural, Relational Universality of Mathematical Knowledge			

Figure 17. The grid of analysis: the connection between the social and epistemological dimensions.

For analyzing the interactive construction processes (chapter 4) the preliminary descriptions of the new knowledge (by students and the teacher) are used as central indicators for the classification of different variants of justifying interactive knowledge construction. In the tension between the two dimensions (the conveyance of rules and factual knowledge) and the construction of new meanings (*communicative* dimension) as well as between the empirical situatedness and the intended relational universality of the new knowledge (*epistemological* dimension), different, specific forms of description and interpretation of the new knowledge, interactively created by students and teacher, will be identified. These „key-words“ of the description of the new knowledge yet to be constructed are analyzed and assigned their places in the classification model. These key-words are evaluated, in the frame of this model of analysis, with regard to the differences and relations between an empirical, situation-specific as well as a theoretical, relational use.

In the episodes, the relations between mathematical „signs / symbols“ and „reference contexts“ (consisting of diagrams and devices of illustration) are in the center of the mathematical activity. As mentioned before, the signs and reference contexts do not represent the new mathematical knowledge directly, but only in a mediating way. Both the (geometrical and arithmetical) diagrams and the (actual) mathematical signs serve as potential carriers of the new knowledge and thus, in this sense, as an externalization of the mathematical knowledge. The mathematical knowledge can be found not only „in the head" of the human being; one can operate with the knowledge by means of the written signs. The purely mental imagination and the exclusively verbal communication about aspects of the number concept represent a first (historical and cognitive) level of development. It is only the encoding of mathematical knowledge by means of signs that makes a decisive progress in the construction of mathematical knowledge possible (cf. Dehaene, 1997, p. 95ff.). The (general) mathematical signs have two essential functions as structural constructions: They are carriers of the knowledge (epistemological dimension), and they become elements of the social communication at the same time (communicative dimension).

The appearance of a seemingly logically fixed meaning of mathematical signs is indispensable for the development of new knowledge in instruction processes and it is necessary for authentic learning of mathematics in the form of a new construction of mathematical knowledge. However, it not only appears in local learning and classroom processes, but is also a general characteristic of modern science and of the social conditions of the construction and acquisition of theoretical knowledge.

> ... only by means of the communication problem the systems character of the knowledge and its operativity ... are pronounced in a pointed and radical sense. Individual-psychologically, a symbolic representation always possesses a certain solid relation with a meaningful imagination. The separation of meaning and symbol only becomes perfect under the pressure of social dynamic. In a static, not changing society resp. in a relatively stable community of scientists the construction of the relation of sign and signified occurs by means of habituation, in a way that the one who grows into such a community acquires the meaning in the use by imitation. Only when this kind of acquisition of meaning is not possible anymore because of the dynamic of the social and scientific development, the communication problem discussed here appears in all its severity. It is the great merit of Hilbert having seen the signs of the time and, by means of the separation of development and justification of knowledge, of sign and signified which he executed, having made room for an idea of concept which allows for variably relating the two separated moments, justification and development, sign and signified, to each other (Jahnke, 1978, p. 164).

OVERVIEW OF THE THIRD CHAPTER

In this book, the third chapter represents the central part, the empirical heart, in which the detailed qualitative analyses of selected instruction episodes are carried out and presented. These analyses require the previously developed theoretical analysis instruments: the epistemological triangle, the communication analysis and later (in chapter 4) the two-dimensional analysis grid.

These theoretical analysis instruments have been „manufactured“ in a specific way and extensively justified in the frame of the elaboration of the fundamental positions on the nature of mathematical knowledge and the role of instructional communication in the first two chapters. Their application to a number of cases (in section 3.3) will bring about further clarification and possible uses of these instruments.

The central concern of this research is the analysis of such instruction interactions, in which the students possibly construct new mathematical knowledge. In order to support this demand of the construction of new knowledge and also the justification of mathematical knowledge in everyday instruction, concepts and materials about substantial mathematical learning environments (section 3.1) have been elaborated and made available to the participating teachers as the basis for small teaching units. The elementary mathematical problems belong on the one hand to the domain of arithmetical structures and patterns, and on the other hand, they relate to geometrical patterns in the domain of figurate numbers. The participating teachers have then prepared the mathematical lessons themselves using this material dealing with a specific topic. In the course of teaching, the students were requested to give explanations and justifications to surprising mathematical particularities. In this way, the children should be encouraged to develop their own justifications and actively make constructions of mathematical knowledge.

The qualitative analyses of the selected teaching episodes are carried out in section 3.3 according to a procedure, which is based on three methodological steps. The first step represents a description of the observed interactive events in the episode. To record the observations besides the videotaped documentary, in particular the extensive and carefully produced transcript of the teaching communication together with the important non-verbal statements is used. This description of the scene and the observable events – made as carefully as possible – is aimed at presenting an orderly, complete picture of the observation, which is not possible by mere presence in the actual course of the teaching. This first step of the analysis allows for gaining an initial overview and organized ordering with the help of „theoretical glasses“ when observing the teaching events.

The two further steps of the analysis are combined in the evaluation presented here. In the frame of the research project, after the extensive survey and the time-consuming transcription of 40 teaching episodes altogether, all the three analysis steps were carried out separately and in detail. This cannot be done here because of reasons of space.

The second step uses methods of the epistemologically oriented analysis of mathematical instruction interactions and tries to examine the epistemology of the mathematical knowledge commonly produced within the interaction. This analysis step uses a kind of a „theoretical magnifying glass“ in order to classify the meaning and epistemological structure of the constructed knowledge more closely. Opposite to a direct observation with ordinary glasses, the magnifying glass helps, through a focus on epistemological problems, to look deeper into the teaching events.

In the third step, the communicative analysis is added. A detailed analysis from an epistemological and communicative view follows, in which the interplay between the two central dimensions „knowledge“ and „communication“ is examined. This analysis can be understood as the use of a „theoretical microscope“ which allows for searching for connections and reciprocal relations in the more indepth details. In this book (section 3.3) the two analysis steps 2 and 3 are combined.

For readers who have very little experience with the very careful qualitative analysis of teaching episodes, and whose previous reflective observations of courses of instruction are maybe not based on a comparable theoretical foundation, as it has been developed in this book, it might be difficult, unusual and demanding to follow at first and to understand the analyses presented in section 3.3. Despite the initial difficulties, it is worthwhile to persist in the careful analyses of the teaching episodes and to gain from them.

A central concern of chapter 3 consists in making visible the broad variety and the wealth of aspects of the mathematical justifications and knowledge constructions in everyday mathematics teaching with the help of the theoretical analysis instruments, as they are expressed in the students' statements as well as in the interactions with the teacher. The analyses show in particular the tension between the situatedness of the mathematical knowledge and the intended possible generalizations. In elementary school instruction, mathematical knowledge is bound to examples and concrete contexts for the students and is not to be directly and completely detached or abstracted from them. However, it is possible for many children to succeed in exceeding the concrete frame of a mathematical example with their own descriptions and interpretations and thus constructing elementary relations and structures. The analyses show the variety and features of the generalizing knowledge constructions described by the children.

CHAPTER 3

EPISTEMOLOGY-ORIENTED ANALYSES OF MATHEMATICAL INTERACTIONS

1. DEVELOPMENT OF MATHEMATICAL KNOWLEDGE IN THE FRAME OF SUBSTANTIAL LEARNING ENVIRONMENTS

The epistemology-oriented approach to qualitative research of mathematical interaction developed in the previous chapters will be now be applied to documented, concrete cases of mathematics teaching (in elementary schools). At the same time, the qualitative analyses will serve the purposes of an additional explanation, further development and differentiation of this approach to research in mathematics education. The analyses will focus on interactively constituted forms of constructing and justifying new mathematical knowledge.

In view of the problem of the development of true mathematical communication in mathematical teaching and learning processes (see section 2.1) it must be stated that (elementary school) mathematics instruction is largely about mediating given, firm rule knowledge for the elementary arithmetical operations. Interactive constructions of actual new knowledge initiated in the instruction process are rather exceptional and require, furthermore, a particular preparation of suitable mathematical learning requirements for the students.

By comparison, the following abilities belong to the very general, essential learning objectives of mathematics teaching: Being able to develop mathematical knowledge and perhaps even to generate really new knowledge in co-operation with other classmates in an autonomous and active way, as well as being able to recognize and to justify the generality of mathematical statements. Initially, these goals should have been already pursued in elementary school by way of examples (see Winter, 1975). On the one hand, the mathematics curriculum for elementary schools (from which the instruction documents employed here are taken) emphasizes the principle of active discovery and social learning in mathematics instruction (see, in particular, Wittmann, 1995).

> A conception in which learning mathematics is conceived as a constructive, discovering process does justice to the tasks and goals of mathematics instruction to a high degree. Thus the lessons must be formed in such a way that the children receive as many occasions as possible for unaided learning in all phases of a learning process (KM, 1985, p. 26).

On the other hand, argumentation skills are particularly emphasized in the list of fundamental competencies to be supported in the curriculum. This is described more closely as

> ... justifying statements, examining declarations, requesting justifications, asking for further information, distinguishing between assumptions and justified statements, fol-

> lowing agreements, applying rules in a consistent manner, uncovering contradictions, not accepting contradictions, accepting counter arguments, ... (KM, 1985, p. 21).

In order to initiate situations supporting active knowledge construction and justification as much as possible in the observed and documented mathematical lessons, the teachers participating in the empirical research were given conceptual proposals related to problem fields within different mathematical topics. Based on these proposals, the teachers themselves developed materials and exercises for short classroom experiments (about 4 to 5 lessons). The fundamental model for the mathematical design of the classroom experiments was the conceptual and concrete practical construction of a substantial mathematical learning environment.

The concept of a mathematical substantial learning environment has been developed in the frame of productive mathematical exercises (cf. Wittmann & Müller, 1990; 1992). According to wide spread opinion, the main purpose of practising in mathematics instruction is the consolidation of knowledge and it consists in the training of skills demonstrated on one or several examples. This conception of practising as merely a „subordinate“ activity which establishes that the new fact information is founded on the assumption that mathematical knowledge is a finished product. „This position conceives of the subject matter as a given „mass“ which only becomes teachable by means of breaking it down into a sequence of single elements. The teacher has the task of „mediating“ these elements to the students in appropriately adjusted fine doses.“ (Wittmann, 1992, p. 175). This mediation of the mathematical subject matter takes place in teaching mainly by way of exercises or problems as mathematical exercises playing a decisive role in shaping attitudes towards mathematical knowledge and in the learning of mathematics (see Bromme, Seeger & Steinbring, 1990, Krainer, 1990).

Traditional mathematics instruction is predominantly concerned with working on exercises. Many mathematics school books consist largely of exercise collections, and the preparation of lessons often concentrates on preparing exercises. The following dangers or possible misuses of working on exercises have been listed by Krainer (Krainer, 1990, p. 341):

- „Monoculture“: Students are given classes of exercises to do, all solvable according to one single scheme, but also controllable. In this way, concepts are not developed any further.
- „Wild growth“: Students are given a colorful palette of exercises, showing neither an objective-oriented guideline nor a common „system“ (an isolated collection of exercises).
- „Misgrowth“: Students are given exercises that have nothing to do with the preceding material, for example with the introduction of a concept, and which can even contradict the concept development.

In this approach I propose that exercises be understood rather as the smallest unit of instructional thinking and acting (von Harten & Steinbring, 1985), as the „building blocks“ of concept development and as the „learning domains“ for certain contents and goals. In order to support this use of exercises and to counteract the dangers and misuses at the same time, it makes sense to develop „systems of exercises“.

An important condition that a construction of a system of exercises should satisfy is the creation of a network that connects the exercises in the system in different ways. Additionally, each exercise should fulfill the functions of motivating, acquiring subject matter, (developing concepts), practising and control of learning. Krainer (1990, p. 344/5) distinguished the following hierarchy for exercise systems:

- microstructure (the single exercises themselves)
- mesostructure (nets of related exercises; groups and sequences of exercises)
- macrostructure (relation between the „environment“ and the „world of theory“)

Even for the practising process, such a coherent system of exercises embodies the idea that it is about a productive engagement with mathematical concepts. Practicing is thus also understood as acquiring new knowledge and not merely as consolidating knowledge information that has already been taught. The intention of this concept is similar to ideas described and developed by other authors: for example, concepts such as the „teaching unit“ (Wittmann, 1984), „structured exercise“ (Wittmann & Müller, 1992), or „substantial learning occasion“ (Krauthausen, 1996). The core idea for practicing in all mathematics teaching has been formulated by Winter as follows:

> ... All... phases contain more or less strong parts of practice and repetition ... it is practiced by way of discovery and discovered by way of practice.... Practising is thus essentially the taking up again of a (discovering) learning process, the re-building of learning situations (Winter, 1984, p. 6).

Against this background of a changed and differentiated view of the meaning and function of exercises in mathematics teaching, Erich Wittmann has worked out the following definition of a mathematical substantial learning environment:

> A substantial learning environment, an SLE, ... is a teaching/learning unit with the following properties:
>
> (1) It represents central objectives, contents and principles of teaching mathematics at a certain level.
>
> (2) It is related to significant mathematical contents, processes and procedures beyond this level, and is a rich source of mathematical activities.
>
> (3) It is flexible and can be adapted to the special conditions of a classroom.
>
> (4) It integrates mathematical, psychological and pedagogical aspects of teaching mathematics, and so it forms a rich field for empirical research.
>
> The concept of an SLE is a very powerful one. It can be used to tackle successfully one of the big issues of mathematics education which has become more and more urgent and which is of crucial importance for the future of mathematics education as a discipline: the issue of theory and practice (Wittmann, 2001, p. 2/3).

Using this fundamental idea of „substantial learning environments“, arithmetic-structural and geometric-visual learning environments have been outlined for the present empirical research project to initiate the active knowledge construction processes of the children. On the one hand, these learning environments contained mathematical problems about „cross-out number squares“ and „number walls“, on the other hand they contained elementary problems about „figurate numbers“ (triangular, rectangular and square numbers). Typical mathematical requirements suggested in these learning contexts represent the so-called *substantial*, i.e. meaningful and relation-rich problems which make many activities and also justifications possible for the young students (cf. Wittmann 1995). Based on the formulated learning environments, the (eleven) teachers participating in the research project prepared teaching materials of their own choice and carried out short teaching units (5 lessons at most) (47 lessons have been documented in 3rd and 4th grade classes in the area of the city of Dortmund).

The fundamental, conceptual ideas of the two arithmetic-structural and geometric-visual learning environments for mathematics teaching in 3rd and 4th grade are outlined below by way of examples.

(1) Learning environment: Operative Structures in Number Walls.

The elementary rule of construction: How is a number wall created?

In the lower row of a four-level number wall (such as the one shown in Fig. 18), four starting numbers have been chosen arbitrarily and written down in the four fields of the row. In order to obtain the numbers in the fields of the row above, two neighboring numbers of the row below are added.

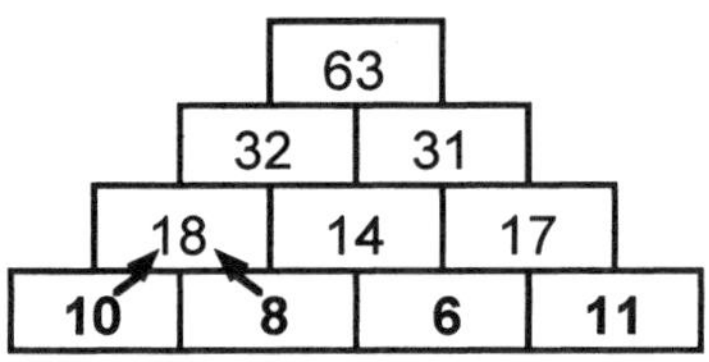

Figure 18. Addition in number walls.

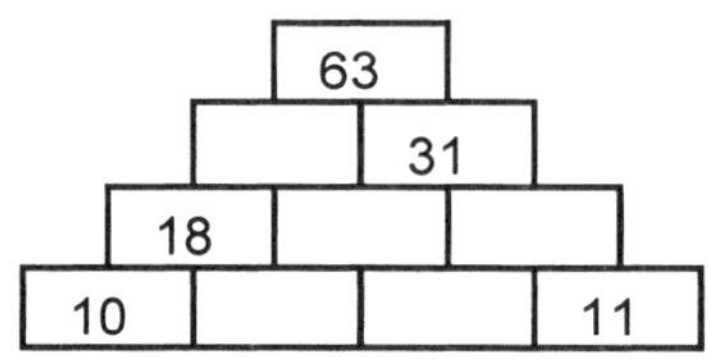

Figure 19. Addition and subtraction.

Thus, 18 results out of the neighboring numbers 10 and 8, 14 out of 8 and 6, and 17 out of 6 and 11. The same procedure is carried out when the numbers of the third row or the highest number, the goal number, are calculated. The essence of the justification is that the value of one lower block has effects on the two blocks above it.

When starting with the exercise the children, in different grades, receive problems within the learning environment of number walls with which they can practise addition and subtraction. The numbers in a number wall are directly connected with one another and this makes the exercises essentially distinct from the isolated addition exercises that can often be found in textbooks. However, a restriction on addition is not necessary; by means of inserting empty blocks, subtraction becomes possible as well (Fig. 19).

Children can discover the first operational connections in number walls by means of „calculating up and down“. This is essentially not about merely calculating further results with given numbers and according to strict rules, but about determining the origin numbers from the given goal numbers by means of trying and systematizing. The following number walls serve as examples for this type of activity (Fig. 20 and 21).

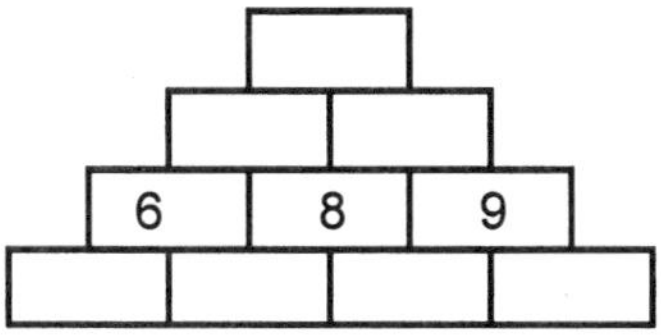

Figure 20. Up and down calculation.

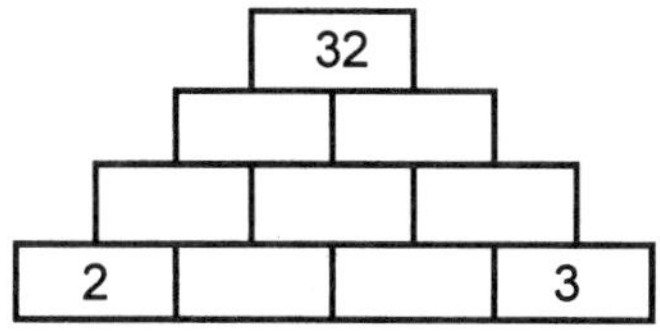

Figure 21. Searching the missing base blocks.

For children, the exercise consists of filling these number walls by using the given numbers and in examining whether a solution really exists or to explore whether several solutions are possible, and in determining all of them by systematic considerations (for example with the number wall in Fig. 21).

The operational structure of number walls can be examined with the help of systematic changes in the base blocks (increase by 1 or by 10): How does the goal block change when one base block is increased by a fixed amount? In a lesson observed as presented in the following, this analysis of the operative structure was carried out with the following number walls (Fig. 22 and 23).

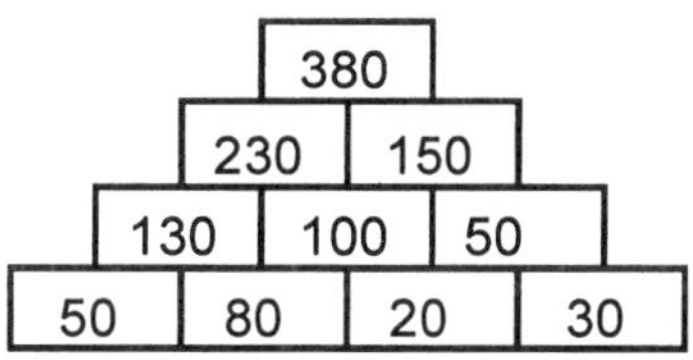

Figure 22. Calculated comparison wall.

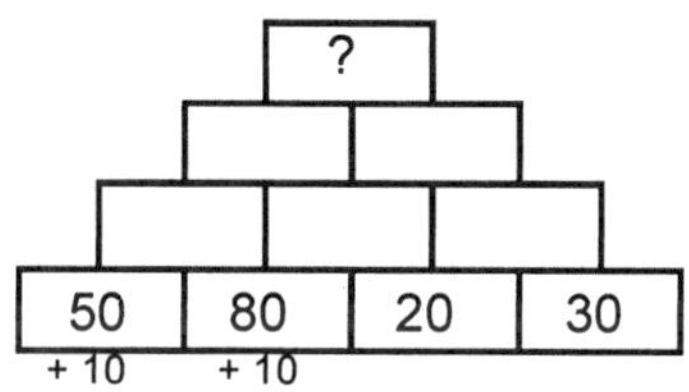

Figure 23. Changes in the arithmetical structure.

After a number wall with the four given base blocks 50, 80, 20 and 30 was completely calculated (cf. Fig. 22), a change by 10 was made first in the left base block, then in a middle base block and it was examined for what effect these changes have on the goal block. A general, functional connection in number walls was uncovered with these example numbers: A change in an outer stone affects the goal block once whereas change in a middle block affects it three times. If an outer block is increased by 10, the goal block increases by 10 as well. Yet if a middle block is changed by 10, the goal block is already changed by 30.

In other documented lessons, a further operational variation of structures in number walls was worked on. This was about the repeated exchange of four given numbers for the four base blocks (cf. Fig. 25).

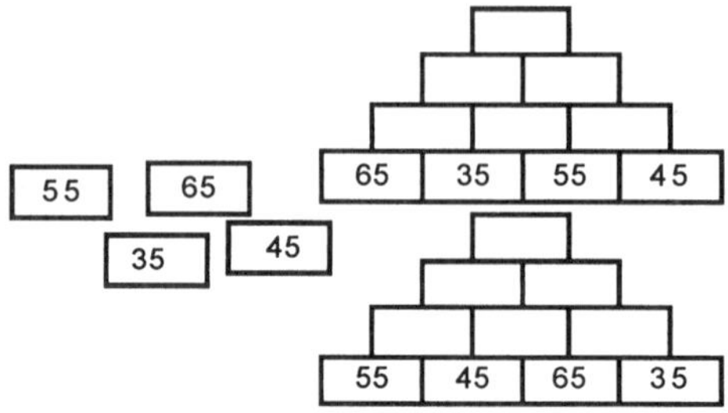

Figure 25. The permutation of the four base stones of a number wall.

This exercise variation in the learning environment „operative structures in number walls“ leads to questions of the following kind: How many different number walls can be constructed in all? How many number walls with different goal blocks can be constructed? What is the connection between the numbers appearing in the goal blocks?

The connected problems for the learning environment of number walls which are presented here by way of examples form, first of all, a substantial learning environment at the level of epistemological knowledge relations. The qualitative analyses of selected classroom episodes which will follow exemplify the integration of mathematical, psychological and pedagogical aspects of teaching and learning mathematics (cf. Wittmann, 2002; in relation with the learning environment of „number walls“, see Krauthausen, 1995; Scherer, 1997; Wittmann & Müller, 1990; 1992; 1996).

[When describing the position of the numbers in a number wall in the course of a transcript or of an analysis, the blocks are numbered consecutively from the bottom to the top and from left to right]

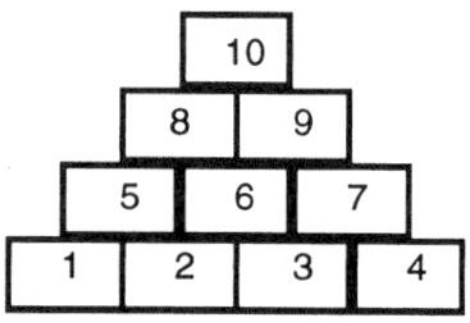

(2) Learning environment: Operative Structures in Cross-out number squares

„Cross-out number squares“ are square arrays of numbers which have been constructed according to a certain principle. In order to be able to solve a cross-out number square, a „crossing out algorithm“ is applied. For this, one number per row and per column is chosen.

For children, the algorithm is couched in the following terms:

1. Choose an arbitrary number and circle it; then cross out all the numbers in the same row and the same column (Fig. 26).
2. Choose an arbitrary, not yet crossed out number and circle it; then cross out all the numbers in the same row and the same column (Fig. 26).
3. Continue the procedure until all numbers are either circled or crossed out.
4. Add all circled numbers.

This addition results in the so-called „crossing sum“: 23 + 13 + 20 = 56.

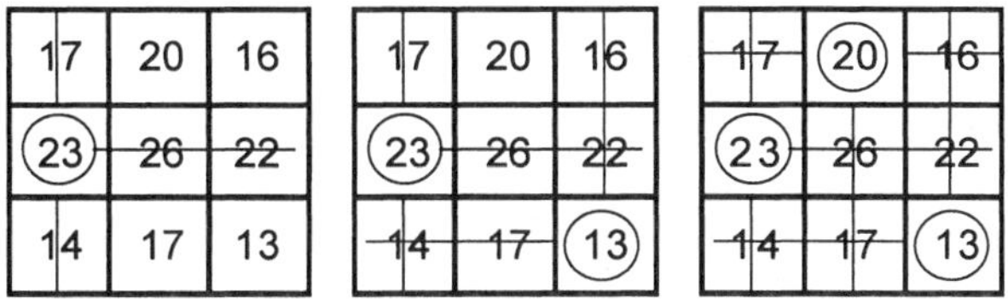

Figure 26. The crossing-out algorithm.

The particular thing about the cross-out number squares is that all children obtain the same crossing sum in the end, independently from their choice of circled numbers (see Fig. 27).

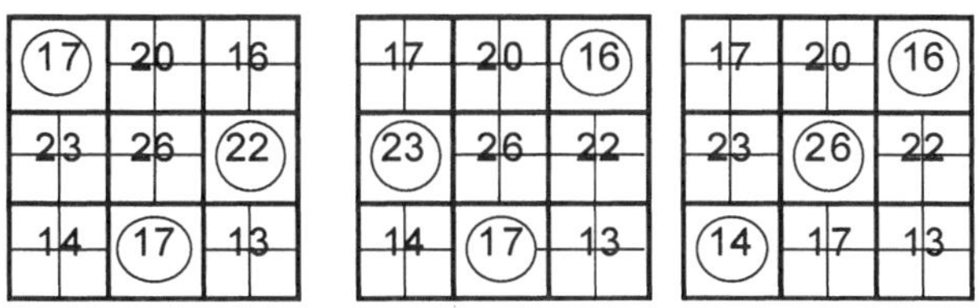

Figure 27. The addition of different circled numbers.

In order to obtain a square with this striking feature, a particular kind of construction is, of course, necessary. A cross-out number square actually represents the inner part of an addition table (Fig. 28).

+	8	11	7
9	17	20	16
15	23	26	22
6	14	17	13

Figure 28. The construction of cross-out number squares from addition tables.

Each inner number is a sum of two border numbers. For the three circled numbers 17, 22 and 17 in Fig. 27, one thus obtains the following additive decomposition into two corresponding border numbers of the table: 17 as 9 + 8, 22 as 15 + 7 and 17 as 6 + 11. This shows that the sum of all six border numbers is equal to the crossing sum. In each row and each column, there is exactly one circled number and each circled number „contains" exactly one number from the border column and one number from the border row, thus the sum of the three respective arbitrarily chosen circled numbers is always constant, namely it is equal to the crossing sum.

This striking feature is the starting point for different exercise problems that could be posed in this learning environment. The introduction into the learning environment of cross-out number squares could be the teacher presenting different number squares at the blackboard which are arbitrarily chosen and have not been constructed according to the principle of cross-out number squares. With such squares, she explains the crossing algorithm, thus the procedure to determine a sum out of three circled numbers. Later she asks the students to look for the biggest or the smallest crossing sums possible.

When the children have understood the crossing algorithm, the teacher writes a true cross-out number square on the blackboard and asks the children to look for the biggest or smallest crossing sum there as well. By means of communication with one another, the children soon discover that they have all obtained the same crossing sum. If this is not the case with one student, there must be a computational mistake. The surprising observation that the procedure always leads to the same crossing sum, provides an opportunity for further exercises and comparisons – which can also be carried out as tasks for partner work or at home.

In the observed lessons on the topic of cross-out number squares, a justification for the constancy of the crossing sum was interactively worked on with children in the frame of the construction of arbitrary cross-out number squares using the border numbers of a table (the sum of the six border numbers is obviously equal to the crossing sum). In order to gain a justification the operational connection between the circled numbers, the border numbers and the crossing sum had to be recognized and varied.

In a substantial learning environment, the following exemplary problems offer excellent possibilities for a further exploration of arithmetical structures in cross-out number squares and for generalizing and finding the patterns of the underlying structures.

Exercise: Can you construct cross-out number squares in such a way that you obtain an arbitrary number as the crossing sum? Construct a 3x3 cross-out number square with the crossing sum 111.

1	2	3	4	5	6	7	8	9	10
11						17	18	19	20
21		23	24	25		27	28	29	30
31		33	34	35		37	38	39	40
41		43	44	45		47	48	49	50
51						57	58	59	60
61	62	63	64	65	66	67	68	69	70
71	72	73	74	75	76	77	78	79	80
81	82	83	84	85	86	87	88	89	90
91	92	93	94	95	96	97	98	99	100

Figure 29. Many 3x3 cross-out number squares in the hundred table.

Exercise: Take a transparent 3x3 frame and put it onto the hundred table (cf. Fig. 29). Look at several 3x3 cuttings out of the hundred table and carry out the crossing algorithm. Do you find striking observations or similarities between the number patterns of different cross-out number squares?

• Examine the connection between the middle number of the 3x3 square in the hundred table and the crossing sum.

• What happens when you move the 3x3 cross-out number square 1 to the right and 1 down?

Exercise: Examine different 4x4 and 5x5 cross-out number squares and their patterns on the hundred table.

Exercise: Calculate the crossing sum of the complete hundred table.

A further possible variation in the learning environment of cross-out number squares consists in working on a „defective" number square with one or more blank fields. The aim of the task is to restore the original cross-out number square by inserting appropriate missing numbers. The following incomplete square represents an easy example (cf. Fig. 30). This was worked on in the classroom of a mixed 3rd and 4th grade class (In this class children of different ages from both grades are all being taught together. This is not the normal situation in German schools but there are at least some experimental schools with classes comprised of two or more grades.) (A qualitative analysis of the teaching episode will be given later). The class developed several different procedures in order to restore the original square.

15	16	17
14	15	16
13	14	

Figure 30. Cross-out number square with a gap.

Cross-out number squares with one or two gaps can be solved easily. With the help of the other numbers, the necessary crossing sum can be determined in spite of the gap. The gaps can be closed using subtraction. When there are more than two gaps, the level of difficulty depends on the position of the gaps. If a combination of three numbers can be unequivocally determined in spite of the gaps, and if the crossing sum can thus be found, the cross-out number square can be solved. Altogether, solutions with up to four gaps are possible. However, children need a certain competence to solve this exercise.

The above examples of problems in the cross-out squares learning environment make it clear how elementary and more advanced exercises are linked in systematically and how they can encourage productive, and thus meaningful and interrelated learning and working. First of all, the epistemological conditions of the elementary mathematical knowledge are generally thematised here as well, in bringing forth the operational structures and general arithmetical connections in the cross-out number squares. The epistemologically-oriented analysis of interaction in episodes of classroom work in the learning environment of cross-out number squares will point to connections with psychological and pedagogical aspects as well (Wittmann, 2002) (for the learning environment „cross-out number squares" cf. Wittmann & Müller, 1990; 1992).

In traditional mathematics instruction in elementary schools, arithmetic problems and exercises were commonly assigned the function of merely training the correct application of computational procedures (cf. Winter, 1982). The substantial learning environments approach does not emphasize the algorithmic procedure of stepwise and correct working out of elementary operations, but the underlying elementary, pre-algebraic structure as an essential foundation for the development of arithmetical connections which are full of relations. Disconnected calculation packages (Wittmann, 1990) do not support such a conception. Productive exercises, such as those presented above, are needed in which an operational connection or an interesting problem contains the essential arithmetical structure of the substantial learning environment.

(3) The Learning Environment of Figurate Numbers: Triangular Numbers, Square Numbers and Rectangular Numbers

The content of this learning environment related to elementary problems in the domain of figurate numbers is based on the relation between geometric configurations of dots and arithmetical determinations of the numbers of dots. It has been observed that the geometric configurations must not be seen simply as direct visual objects. „Visualizations in mathematics are not pictures, not pictorial expressions or illustrations" (Otte, 1984, p. 10). These geometric diagrams are carriers of multiple structures which have yet to be interpreted by the students.

One phase of the introduction into triangular, rectangular and square numbers is about exploring the corresponding geometric continuation principles by means of a small quantity of presented configurations, i.e., about inferring general connections from the few given configurations. This geometric continuity is supposed to be described by means of a sequence of numbers, as in arithmetical equations and in this way, making visible the continuation principles on the arithmetical side (see below).

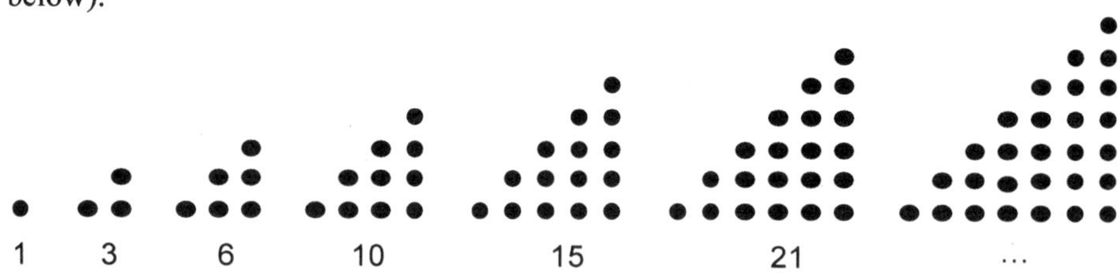

Figure 31. Triangular numbers.

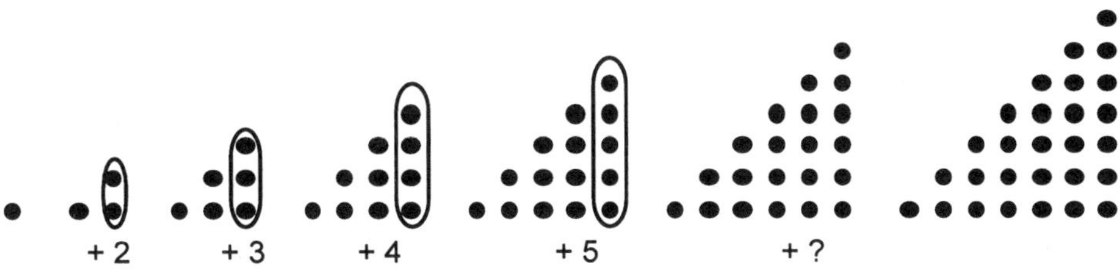

Figure 32. A possible continuation principle.

A possible arithmetical continuation structure on the basis of the geometric structure:

1 = 1
1 + 2 = 3
1 + 2 + 3 = 6
1 + 2 + 3 + 4 = 10
1 + 2 + 3 + 4 + 5 = 15
1 + 2 + 3 + 4 + 5 + 6 = 21

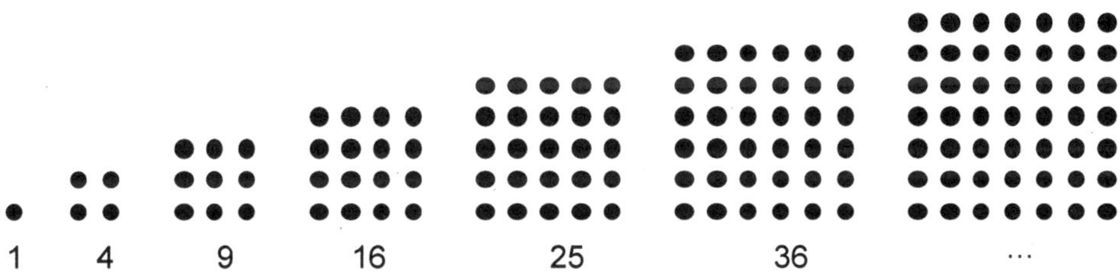

Figure 33. Square numbers.

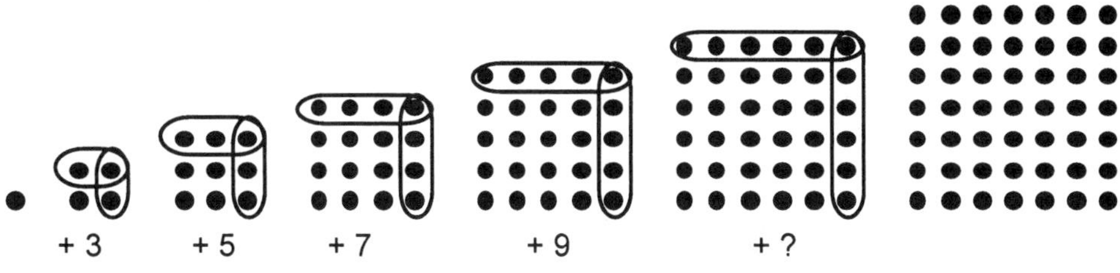

Figure 34. A possible geometric continuation principle.

A possible arithmetical continuation structure, on the basis of the geometric structure:

$1 = 1$	$1 \cdot 1 = 1$
$1 + 3 = 4$	$2 \cdot 2 = 4$
$1 + 3 + 5 = 9$	$3 \cdot 3 = 9$
$1 + 3 + 5 + 7 = 16$	$4 \cdot 4 = 16$
$1 + 3 + 5 + 7 + 9 = 25$	$5 \cdot 5 = 25$
$1 + 3 + 5 + 7 + 9 + 11 = 36$	$6 \cdot 6 = 36$
...	...

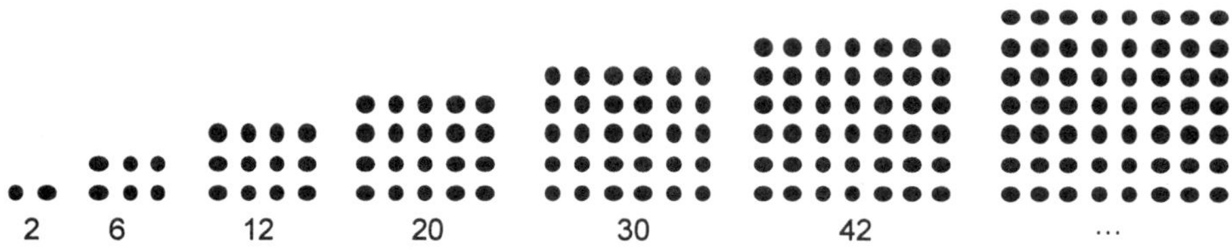

Figure 35. Rectangular numbers.

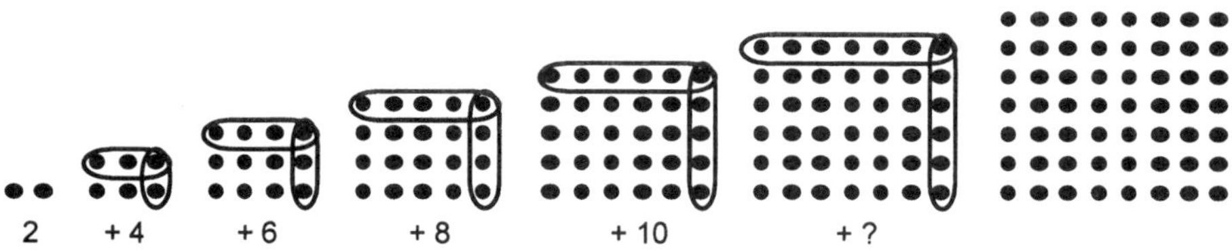

Figure 36. A possible geometric continuation principle.

A possible arithmetical continuation structure, on the basis of the geometric structure:

$2 = 2$	$1 \cdot 2 = 2$
$2 + 4 = 6$	$2 \cdot 3 = 6$
$2 + 4 + 6 = 12$	$3 \cdot 4 = 12$
$2 + 4 + 6 + 8 = 20$	$4 \cdot 5 = 20$
$2 + 4 + 6 + 8 + 10 = 30$	$5 \cdot 6 = 30$
$2 + 4 + 6 + 8 + 10 + 12 = 42$	$6 \cdot 7 = 42$
...	...

In the frame of the introduction into the geometric and arithmetical continuation principles of these elementary figurate numbers the following problems can be posed, for example:

Exercise: Find the 20th (100th) triangular number.
Exercise: Find the 20th (100th) square number.
Exercise: Find the 20th (100th) rectangular number.

Exercise: Compare the square numbers with the corresponding rectangular numbers. What do you observe? Justify your observations with the help of the geometric representation.

Exercise: Compare the triangular numbers with the corresponding rectangular numbers. What do you observe? Justify your observations with the help of the geometric representation.

Exercise: Compare the triangular numbers with the corresponding square numbers! What do you observe? Justify your observations with the help of the geometric representation.

These exercises open a second working phase in the „easy figurate numbers" learning environment, focused on the comparison of different configurations in which, for instance, chip configurations can be divided. In the following, only a few dissections are shown as possible examples which are meant to illustrate how one can reciprocally discover other figurations in chip configurations.

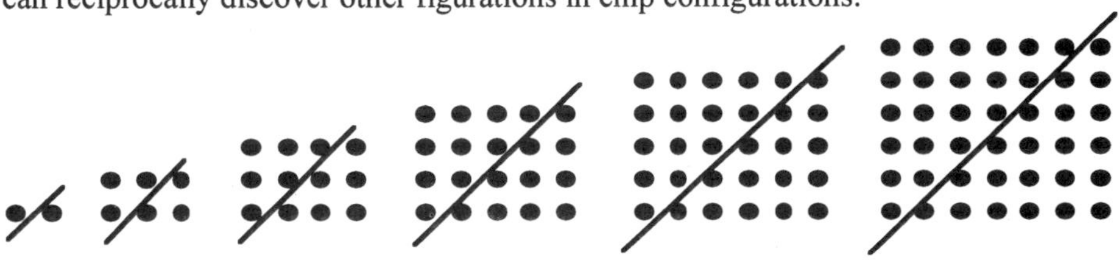

Figure 37. Two equal triangular numbers in rectangular numbers.

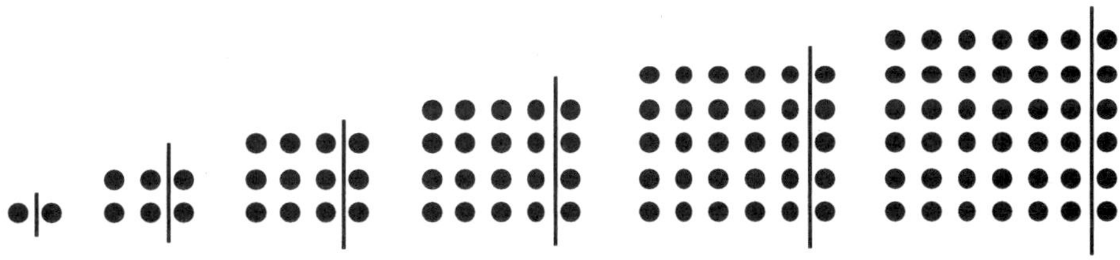

Figure 38. Square numbers in rectangular numbers.

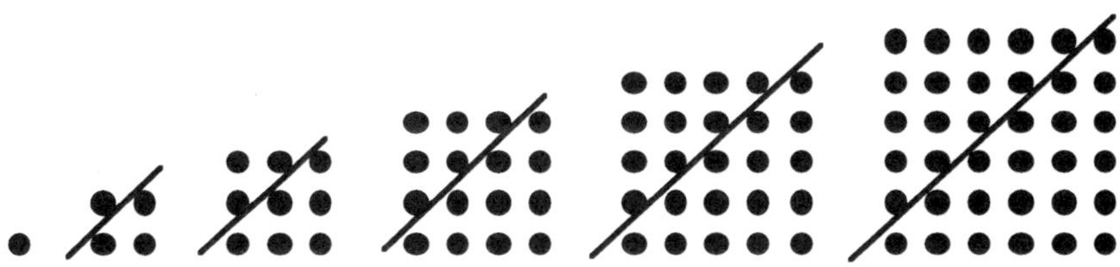

Figure 39. Two successive triangular numbers in square numbers.

Some problems consist of the construction and justification of geometric dissections and compositions of chip configurations. Other problems aim at further development of arithmetical structures or elementary equations. For example, a quick arithmetical procedure for determining big triangular numbers can be developed.

In a third phase one could concentrate, for example, on the examination of the number of the border chips in elementary figurate numbers. In the following, an example of this kind of problem is given.

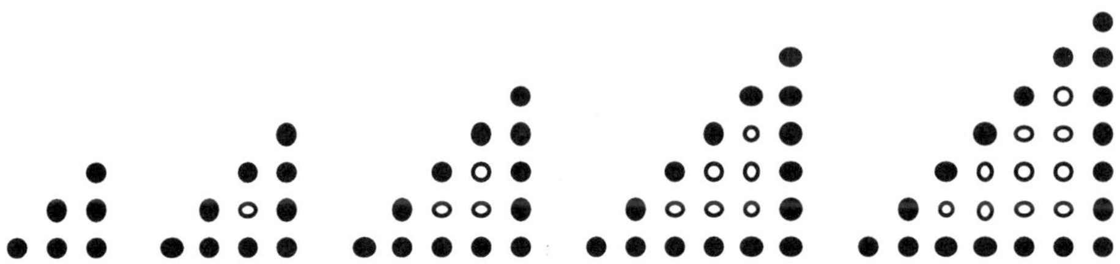

Figure 40. The „border“ of triangular patterns.

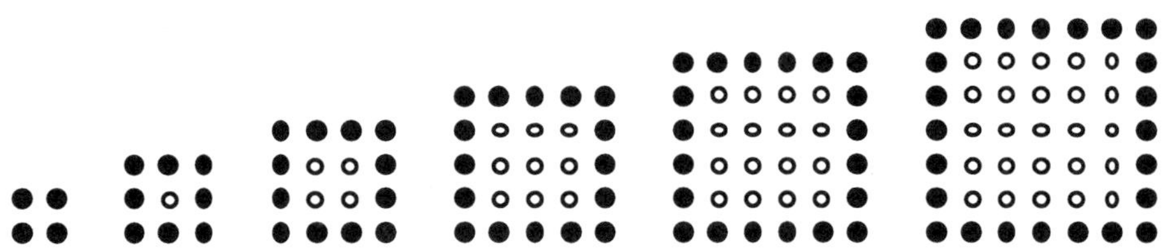

Figure 41. The „border“ of square patterns.

In this context of problems, one can work again on the valid geometrical and arithmetical progression principles. Furthermore, figurations of a similar kind can be discovered in the continuation of the figurations in the inner part of the border; this geometric pattern can thus be employed for the development of a quick arithmetical procedure for determining the border of a large triangular or square number.

In the frame of the geometric-visual learning environment of elementary figurate numbers, knowledge relations and structures can be made evident on two levels: on the arithmetical level of numbers and on the visual geometric level. Here, generalizations can be justified, for example, by means of geometric visualizations, such as, by translating the relations and the structure of a dot pattern representation into an arithmetical relation. However, the visual level usually allows several interpretations whereas numbers tend to be relatively unambiguous.

Visualizations and illustrative materials play an important role in the teaching and learning of mathematical knowledge at the elementary school level. Attention must be paid to the fact that these visualizations do not work automatically, but have to be actively interpreted and structured by the child.

> There is no direct way from the means of visualization to the student's thinking, at the best different difficult detours. The feature of the number 3 is not visible in three smarties or three Lego bars, as if it could be inferred by the child by means of mere contemplative observation. It is an abstraction which yet … is not successful by means of mere leaving out the supposedly unimportant (Lorenz, 1995, p. 10).

This abstraction must be ultimately achieved by the child itself, by means of „reading" new – yet invisible – relations and structures into the visualizations. In this way the child is constructing new knowledge.

2. EXAMPLES OF TEACHING EPISODES AS TYPICAL CASES OF THE EPISTEMOLOGY-ORIENTED INTERACTION ANALYSIS

For teachers participating in the research project, the conceptual suggestions for mathematical learning environments formed the basis of their own planning and preparation of a short instruction sequence which was then observed and documented. However, it must be emphasized that the „conceptual suggestions for learning environments" were not intended for a direct use in teaching practice. The practical realization of the learning environments into an instructional plan and concrete lessons had to been performed by each teacher themselves. Here, Wittmann's remark in his description of the mathematics education developmental research carried out in the frame of the project „mathe 2000", is effective without restriction:

> In order to avoid misunderstandings, it must be pointed out that ‚design', according to our understanding, does not directly, but only indirectly refers to teaching and learning. The learning environments developed by us leave the teachers and students room for the productive realization, for their own initiative and self-organized learning, and ask for this (Wittmann, 1997, p. 45).

The epistemology-oriented analyses of classroom processes of interactive constructions of mathematical knowledge presented in the following represent a selection of a few documented classroom episodes from an extensive collection of videographed lessons. In the underlying research project (cf. Steinbring 2000), 11 teachers have participated in short teaching sequences of about 5 lessons each. Altogether, 47 mathematics lessons have been videotaped. From this extensive data material, 13 longer episodes of 15 minutes on the average have been selected and comprehensively analyzed in three steps according to the procedure described in section 1.2.

In the *first step,* which prepares the actual analysis, a „description of the particular episode according to the single phases" is made (the tabular overview over the phase division of each single episode precedes the corresponding transcript). This first step of the analysis allows a first overview and organized ordering with the help of „theoretical glasses" when observing the teaching events.

The actual epistemological analysis starts with the *second step of analysis.* In this step, an attempt is made to reconstruct the relations between the „signs / symbols" and the „objects / reference contexts" which were interactively constructed in the course of the mathematical communication within the episode and thus to reconstruct the „development of the epistemological interpretation in the course of the interaction". In order to go deeper under the visible surface of the observed events in mathematics teaching, „theoretical glasses" are required, which allow for recognizing important differences and details.

In the procedure which was applied in the course of the analysis, the *third step* consists in a „detailed analysis of patterns of justification from a communicative and an epistemological point of view". However, this step of analysis no longer refers to the complete episode, but concentrates solely on the single sequences of construction and justification of mathematical knowledge that could be identified within the episode, in order to reconstruct the different patterns of (new) mathematical knowledge in them. This analysis can be compared with the use of a „theoretical microscope“.

The following application of the epistemology-oriented interaction analysis relies on a few, specifically selected, typical interaction episodes and it also uses the detailed analyses which were carried out according to the comprehensive methodical procedure. Yet the analyses presented here will not be toally detailed, but will concentrate on essential features and fundamental aspects of the epistemology-oriented interaction research. As already mentioned before the two analysis steps 2 and 3 are combined in the presentation of the qualitative analyses in section 3.3..

The selected episodes originate from the instruction of three female and one male teacher. The content of the instruction sequence prepared and carried out by teacher A (female) was based on the „cross-out number squares“ learning environment. In teacher A‘s class there were about the same numbers of 3rd and 4th grade students. On the one hand, there were four tables for children working in small groups (of about six) at which the students had their more or less „fixed“ seats in the classroom. On the other hand, there was also one big „plenary table“ directly in front of the blackboard and separated from the group tables by book shelves; whole class interaction phases usually took place at the large table.

Teacher B (female) chose problems from the „number walls“ learning environment. At the time of the classroom documentation, her class was in the first semester of grade 4. In the classroom, children sat at five group tables with four to six students at each table. Teacher B, who usually worked as a teacher educator in a study seminar, had not taught this class continuously, but only sporadically as a subject teacher. Yet the children already knew the teacher quite well from her earlier teaching assignments in the mathematics instruction of this class.

Further episodes analyzed here originate from the instruction of teacher C (male) and teacher D (female). Both prepared their instruction sequences based on exercises and problems from the „figurate numbers“ learning environment, but their instruction focused on different aspects of the environment.

The instruction in teacher C‘s class was recorded in a small town elementary school. For the most part, the phases of common work in this 3rd grade class took place in a so-called „theater circle“ in which all students assembled on small benches in front of the black board. For the phases of single and group work, however, the children remained seated at the five tables scattered around the classroom.

Teacher D‘s class was a 4th grade class. The instruction that was carried out for the project in this class did not take part in the usual classroom , but in a room that was made available by the head of the school especially for the recording.

3. ANALYSES OF TEACHING EPISODES FROM AN EPISTEMOLOGICAL AND COMMUNICATIONAL PERSPECTIVE

3.1 Teaching Episodes from Teacher A's Instruction

3.1.1 Analysisof the First Scene from the Episode „Why is the magic number 66 always obtained?"

This scene has been taken from the second lesson of the teaching sequence carried out by teacher A on the topic of „cross-out number squares". At the beginning of this lesson, the students have applied the crossing algorithm to a particular 3x3 cross-out number square several times. This time, no different crossing sums resulted – as was the case with the number square they had worked on the day before – the crossing sum was always the same. In this class, children often spoke of magic squares and of magic numbers instead of cross-out number squares and crossing sums.

Magic formula 1:

Calculate the table!

+	13	9	4
10			
12			
18			

Compare to your magic square!

Figure 42. Page 1.

Magic formula 2:

Paint the circled numbers of your magic square red!

23	19	14
25	21	16
31	27	22

Figure 43. Page 2.

Magic formula 3:

Write the red-painted numbers into the circles. Find the according addition exercises and write them into the boxes!

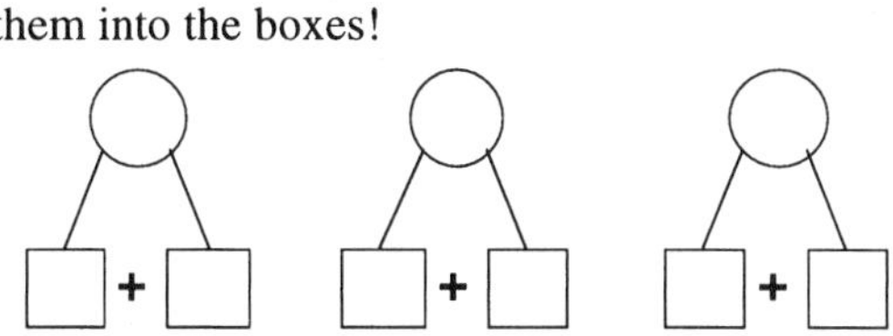

Figure 44. Page 3.

In the group work phase, children were supposed to discover and understand the „trick“ behind this phenomenon. The teacher had prepared a copy-book for each child with three so-called „magic formulas“ that served as working material and a basis for trying to detect the trick. The above illustrations show the three pages of this copy-book (Figs. 42, 43 & 44).

Children were supposed to work in pairs on these three pages consecutively. In the course of this phase, students were coming to the teacher alone or in pairs and were trying to give justifications, or they asked for explanations for different exercises in their copy-books. During this instruction phase, the student Kim comes to the teacher and explaines her justification.

10 Ki [*shows T the first page of her copy-book with the magic formulas and explains the solution*] Mhm, we know it now-, now. One divides the six-, one can divide the sixty-six into three different things

11 T Yes.

12 Ki Always. And one can do that very many times.

13 T Yes.

14 Ki So that there are nine solutions. And if one, like, takes the twenty-two, the twenty-one,andthe whatever, then the result is always sixty-six. Or if one takes the thirty-one, the sixteen and the nine-, nineteen, then the result is sixty-six <u>again</u>. [*Ki, her partner and the teacher appear on screen*]

15 T Yes. Ehm. ... That‘s right, Kim. You found that out right [*T points at the sheet*]. Now it is important, ehm, why [*T looks at Ki who turns away from T*] the result is always sixty-six and not, for instance, seventy-seven or one hundred or

Magic formula 1:

Calculate the table!

+	13	9	4
10	23	19	14
12	25	21	16
18	31	27	22

Compare to your magic square!

Figure 45. The filled square.

16 Ki Yes, because there-, [*points at the sheet*] these are not the numbers there-, these are no numbers [*points at the sheet again*] which one has divided by that, with the seventy-seven. [*3 sec break - T is turning her hand doubtingly*]

17 S1 [*addresses the T from the side*] Mrs Lange [*T does not pay attention to him*]

18 Ki These are not the three-, all numbers [*points at the sheet*], which one can divide by seventy-seven.

19 T Yes, but you can't divide those by sixty-six either [*points at the sheet*] That works

20 Ki Yes okay, okay. [*Ki points at the sheet with her finger*] But these are all numbers which one-, well, these are not all numbers from sixty-six divided by three. But if, yes, [*Ki points at the sheet again*] if one had seventy-seven now, one would have to put fifty-two here, for example. Then one would have to put five here, and then yet twenty.

21 S1 [*is now standing at the other side of the T*] Mrs Lange [*incomprehensible*]

22 T [*takes Kim's magic formula copy-book from the table into her hand*] Mhm? [*to S1*] Is it too much? Just put it there. [*while S1 puts a copy-book with magic formulas on the table of the T, the T turns to Kim again*] Ehm, Kim, consider exactly once again how these numbers here emerged [*points at Kim's copy-book*] and then consider once again, how that-, where you find the sixty-six. Go through the magic formulas once again, maybe you'll find it out.

This episode (10 – 22) with the topic „Kim imagines three circled numbers as ‚three parts' of the magic number 66" contains two phases and can be summarized as follows.

Phase 2.1 (10 – 15) Kim explains the numbers in the square as nine „things" out of the number 66.

Kim formulates a justification according to which she imagines that the 66 is divided into three different „things" (numbers). One could do this several times until one has nine „things". Then she says: „And if one, like, takes the twenty-two, the twenty-one, and the whatever, then the result is always sixty-six. Or if one takes the thirty-one, the sixteen and the nine-, nineteen, then the result is sixty-six again." (14). The teacher agrees, but she also asks why the result is not 77 or 100.

Phase 2.2 (15 – 22) Kim explains how the numbers in the square ought to be constituted for the magic number 77.

The numbers in the square have – according to Kim – not resulted from dividing 77. One cannot divide by 66 or 77, as the teacher says. Kim corrects herself: „Yes okay, okay. [*Ki points at the sheet with her finger*] But these are all numbers which one-, well, these are not all numbers from 66 divided by three." She adds an explanation which numbers could – for instance – come into question for the 77. „But if, yes, [*Ki points at the sheet again*] if one had 77 now, one would have to put 52 here, for example. Then one would have to put 5 there, and then yet 20." (20). The teacher asks Kim to consider once again how the numbers in the square emerged (22).

Following this paraphrasing description, I will now proceed to obtain insights into this scene by way of a communicative and an epistemological analysis. The communicative analysis emphasizes the semiotic function of the interactively produced

signs with the help of the relation „signifier – signified". In the epistemological analysis, the particularities of mathematical signs are to be examined. Thus it will be especially demonstrated in which sense the communicative signs observable in the mathematical interaction are also mathematical signs.
The following semiotic characterization attempts to illustrate the relation between the signifiers and the signifieds in this interaction. The focus is on extensions or differentiations of signifiers to the same signified, yet progressive chains of a continuous change of signifier and signified are also possible (cf. Presmeg 2001).

With the first signifier, Kim refers to the calculated cross-out number square together with the statement that 66 is the magic number.

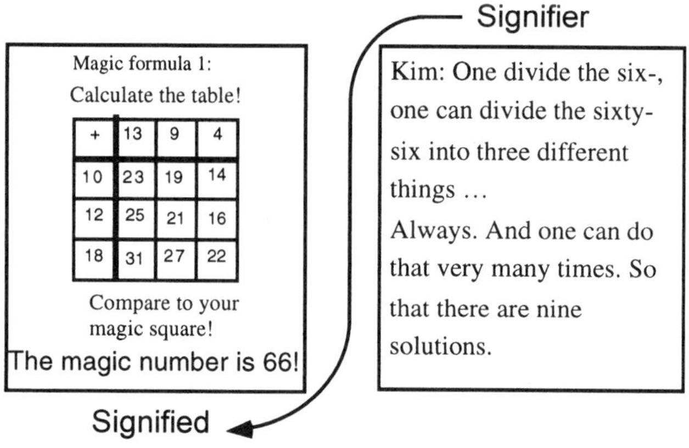

Figure 46. The semiotic model: Magic number and magic square.

In Kim's contribution, the signified communicated by her can be essentially summarized as follows: „66 can be divided into three different things several times, until one has nine solutions". With reference to the signified „calculated number square with crossing sum 66", Kim constructs a new semiotic sign as a unity of signifier and signified. So far, the interpretation of the magic number was of the kind that it was inferred from the addition of three circled numbers as the result. But now, a new interpretation of the magic number is intentionally postulated from Kim's contribution. The magic number can be dissected into three numbers several times, until one obtains nine numbers and places these into the nine fields of the square. The new semiotic sign is thus a new interpretation of the role of the magic number. The idea of the magic number as a result of calculation refers back to the idea that the magic number can be taken as the starting point of the construction of a cross-out number square with exactly this magic number. This intention is concretized by means of example numbers.

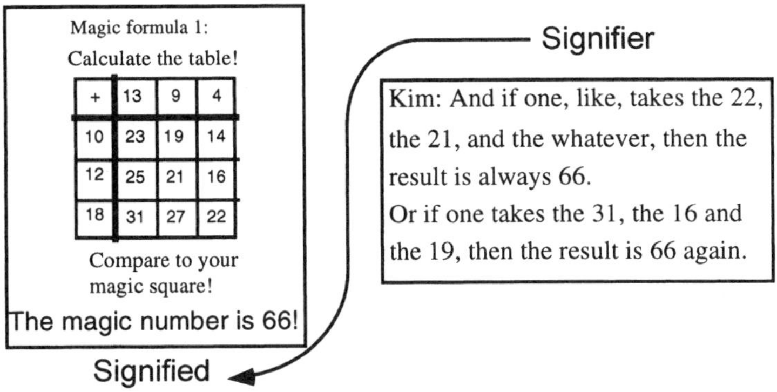

Figure 47. The semiotic model: Magic number and magic square.

On the basis of what was said before, the example numbers 22, 21 and „whatever" or 31, 16 and 19 in the signifier formulated in this way contain the interpretation that they have developed out of a dissection of the 66 in such a way that their sum must be exactly 66. This contribution made by Kim can be regarded as a concretizing confirmation of her interpretation of the new role of the magic number. That this intentional interpretation of taking the magic number as the starting point of the construction of cross-out number squares is true becomes obvious in the interaction at the point where the teacher asks why 77 is not the magic number. Thereupon, Kim develops a first construction step for a magic square with the magic number in a hypothetical way.

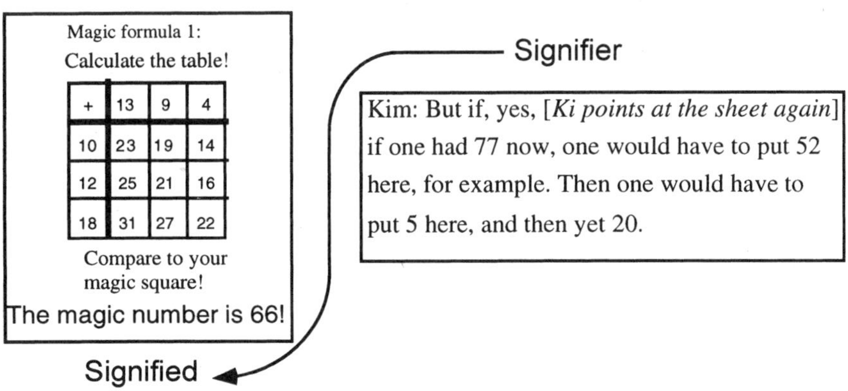

Figure 48. The semiotic model: magic number and magic square.

With reference to the structure of the present cross-out number square with its given numbers, Kim constructs three other possible numbers which could be imagined as circled numbers and whose sum would result in 77 exactly with this signifier. By means of the indicating reference „... put there...", possible positions of these circled numbers in the square are hinted at. With this signifier, Kim illustrates at the „counterexample“ formulated by the teacher how her knowledge construction is to function. The magic number is dissected into three numbers, and these numbers are written into the positions for three circled numbers in the square. This is continued until the square is filled with numbers constructed in this way.

From an epistemological point of view, it becomes clear here that the fundamental idea of the „inversion“ of the „calculation of the magic number from the three circled numbers in the square“ to a „dissection of the magic number into three numbers which are used for the construction of the square“ is of a genuinely mathematical nature. However, attention must be paid to the fact that this suggested construction cannot be concretely realized so easily; the successful production of a cross-out number square from a given crossing sum must use the 6 border numbers (cf. the analysis of the following teaching episode).

Kim constructs the new signifier, and thus a new relation to the signified, by means of verbal descriptions and of demonstrating. This relation can at the same time be interpreted as a mathematical relation, as it embodies relational connections in the cross-out number square and needs the presupposed conceptual, arithmetical structure of the context of cross-out number squares or producing this relation between the signifier and the signified.

The particular construction that Kim presented for the specific number square by means of the triple dissection of the 66 into a sum out of three terms in order to put the nine terms into the nine fields of the square represents an „inversion“: The magic number is not a „subsequent“ result of a calculation rule. It is put at the beginning of the construction of the number square. Thus the sum of three circled numbers is always „66“. The „equation“ $\bigcirc + \bigcirc + \bigcirc = 66$ was only understood as a rule to calculate the magic number so far. But Kim interprets this arithmetical connection now in the way that the circled numbers can be the result of a decomposition of the number „66“ into three numbers. With her proposal, Kim constructs a new mathematical relation: The dissection of the magic number into three terms of a sum makes the construction of a particular number square possible.

In this way, the particular knowledge construction produced by Kim with the help of verbal descriptions and showing becomes a true mathematical construction. It represents embodiments of arithmetical structures in the cross-out number square and it relies on the presupposed arithmetical rules of the cross-out number square. With the help of the epistemological triangle, this mediation between „sign / symbol“ and „object / reference context“ can be modeled as an interactive, mathematical knowledge construction by using a particular example (Fig. 49).

In the following interaction, Kim‘s considerations are further illustrated. The teacher asks why the crossing sum is always 66 and not 77 or 100. Kim argues that the numbers in the given square did not develop out of a dissection of the 77. She says: „These are not the three-, all numbers [*points at the sheet*], which one can divide by seventy-seven.“ (18). Here, she speaks imprecisely „... divide by 77 ...“, but means the correct interpretation: dissection of the 77 into three terms of a sum. Her intention becomes clear in her „correction“ which she formulates to the teacher‘s objection. „But these are all numbers which one-, well, these are not all numbers from sixty-six divided by three.“ (20) One can assume that Kim is talking about a dissection of the 66 into three numbers when she says imprecisely „divided by three". Furthermore, what is most likely intended here is that the numbers in the given square have been developed from 66, and thus cannot result in the sum of 77.

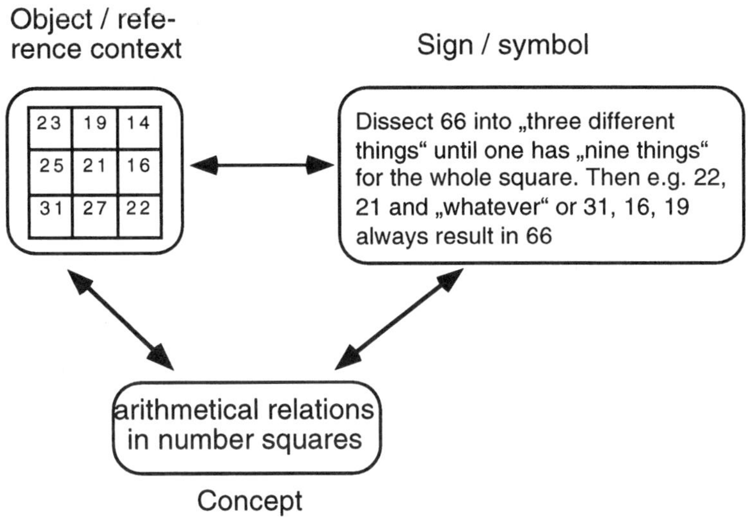

Figure 49. The epistemological triangle: Magic number and magic square.

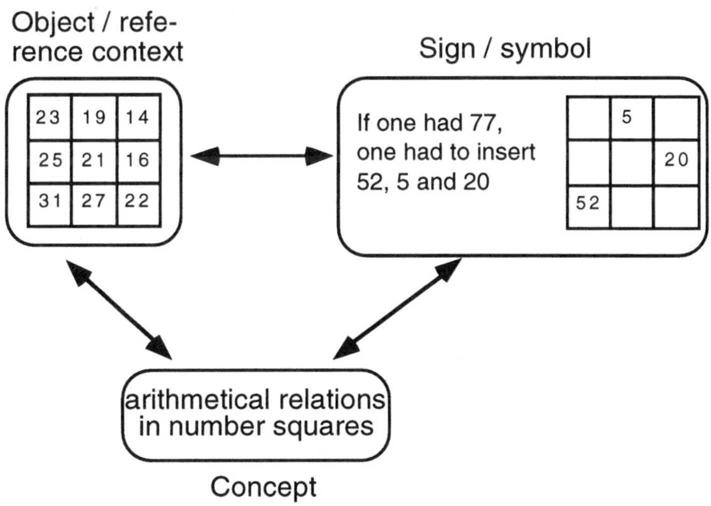

Figure 50. The epistemological triangle: Magic number and magic square.

Then Kim adds as to how the numbers in the square ought to be constructed if the crossing sum were 77. „But if, yes, [*Ki points at the sheet again*] if one had seventy-seven now, one would have to put fifty-two here, for example. Then one would have to put five here, and then yet twenty.“ (20). The knowledge construction can be described in the epistemological triangle (Fig. 50).

In the course of this phase, Kim develops her argument in partial steps. It becomes obvious that she presents the construction of the particular number square – which always results in the same crossing sum – approaching it from the crossing sum: The square emerges by means of dissecting a number (66 or later 77) into sums of three terms and inserting these terms into the nine fields of the square.

3.1.2 Analysis of the Second Scene from the Episode „Why is the magic number 66 always obtained?“

As for the preceding scene, this scene also originates from the second lesson taught by teacher A. In the meantime the common discussion of the mathematical problems worked on during the individual and pair work has started. Some students remarked that the sum of the six border numbers results in the magic number 66.

+	13	9	4
10	23	19	14
12	25	21	16
18	31	27	22

Figure 51. The sum of the six border numbers results in the magic number 66.

In the following, three central statements by Judith from a longer phase of interaction between Judith, the teacher and other students will be selected. Step by step, Judith‘s contributions develop a new knowledge relation in cross-out number squares between the border numbers and the three circled numbers in the square. The teacher says that she is not yet satisfied with the remark that the sum of the six border numbers results in the magic number. She calls Judith who goes to the black board on which the calculated cross-out number square (see Fig. 51) can be seen.

352 J: Yes, like this, well I could also draw freehand <u>another</u> magic square on the blackboard now. <u>Because</u> now, um, I‘d say, these would all be very, very different numbers [*points at the border numbers]*. Because if one, um, … um, then always, um, it was <u>ten</u> plus <u>thirteen</u> is, eh, is <u>twenty-three</u> [*points accordingly at the mentioned numbers*]. Yes and that , … and then if one then circles these now [*points at the middle numbers*]. So an…, and then, so and then there is the <u>twelve</u> once, the <u>ten</u> once, the <u>eighteen</u> once,

the thirteen once, once, the nine once and the four is in the game once. [*points accordingly at the numbers she mentions*]

353 T: So, and we want to make that clear now. You are exactly on the right track here. Judith .... Judith, can you show that? [*points at the whole number square*]. Ehm, you say, each number is only in the game once.

Judith points at the border numbers (Fig. 51) and says: „Yes, like this, yes, well, I could draw freehand another magic square on the blackboard now. Because now, um, I'd say, these would all be very, very different numbers [*points at the border numbers*]" (352). In this way, Judith expresses a generality or variability in the particular case of the present concrete cross-out number square with given border numbers. Then she wants to introduce in her argument the construction principle, according to which the numbers in the cross-out number square develop from the border numbers. This is no longer done „generally", but by means of using the given numbers. However, Judith's intention of generality in speaking about the concrete case is quite obvious here. She says: „Because if one, um, ...um, then always, um, it was ten plus thirteen is, eh, is twenty-three [*points to the numbers she mentions*]" (352). One could say that, for want of a general *designation*, e.g. with algebraic forms of notation, she takes the numbers already present as „fill-ins" for the addition structure of two border numbers to construct a number in the cross-out number square.

Judith continues her argument in the following way. She assumes that the numbers in the diagonal were circled (that they were the selected crossing numbers): „... and then if one then circles these now [*points at the middle numbers*]" (352). She adds that then each of the six border numbers is „in the game once", i.e. appears in the additive construction of the three circled numbers: „So an..., and then, ...so and then there is the twelve once, the ten once, the eighteen once, the thirteen once, on-, the nine once and the four is in the game once. [*points at the numbers she mentions*]" (352).

The teacher confirms Judith's justification and tells her that she is on the right track. However, she wants to clarify this procedure. In several steps, Judith's construction is explained „exactly". Judith first circles the 21 and crosses out the numbers in the corresponding column and row: 19, 27, 16 and 25. Then she circles the 14 and accordingly crosses out the numbers 22 and 29. The dissection exercises are written down next to the cross-out number square: 12 + 9 = 27 and 10 + 4 = 14 (see Fig. 52). The circled numbers and the respective border numbers are marked with the same color. Finally, Judith circles the number 31.

364 J Well here ten plus four, we also have them in the game ... [*writes on the blackboard: 10 + 4 = 14*]. Yes. And then, um, I also circle these two, yes also the thirty-one. So [*gets another pencil from the teacher and circles the number 31*]. And then I also have the eighteen and the thirteen in the, yes, in the game [*marks the border numbers 18 and 13 with the pencil*]. So and none is in

there twice, because I have also crossed out some of them [*writes onto the blackboard: 18 + 13 = 31*].

12 + 9 = 21
10 + 4 = 14
18 + 13 = 31

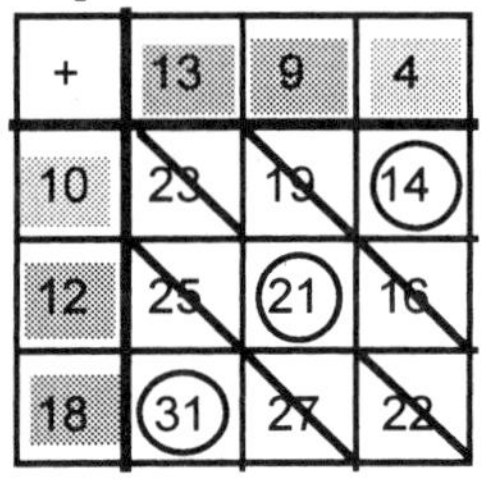

Figure 52. Border numbers and circled numbers.

Judith says: „And then I also have the eighteen and the thirteen in the, yes, in the game [*marks the border numbers 18 and 13 with the pencil*]. So and none is in there twice, because I have also crossed out some of them ...“ (364). Judith remarks explicitly that no border number appears more than once („... none is in there twice...“) and gives a justifying argument for that („... because I have also crossed out some of them...“). The third addition exercise is written onto the blackboard, resulting in the blackboard picture shown above (Fig. 52). Judith explained the connection between the border numbers and a certain combination of three crossing numbers with a tendency towards the „general“, thus in a way which, in principle, already contains the *general ideas*.

Later Judith formulates once more the idea that each border number appears exactly once as a summand in a circled number.

396 J Well. Yes, well, if you calculate there eighteen plus twelve plus ten plus thirteen plus nine plus four, now, then it is always sixty-six. Yes, um, now if you add all these numbers together then. We have the thirty-one, that, that yes, ... yes that is the same as the eighteen and the thirteen. Well, um, and, and that from the plus numbers, thus from the border numbers that yes, um, eh, ... now each border number, that there comes once into the game, um, if one crosses something out, or so. That, that yes, and that‘s why one is allowed in one row to um, now then ... yes then circle only one number. Yes, because otherwise, one number would be contained twice.

First, Judith states once more that the sum of the border numbers is 66. „Well. Yes, well, if you calculate there eighteen plus twelve plus ten plus thirteen plus nine plus four, now, then it is always sixty-six. Yes, um, now if you add all these numbers together then.“ (396). She takes concrete numbers, but in the second part of her statement – „... if you add all these numbers together then“ – she refers to the fact that she already generally intends that the sum of the border numbers results in the magic number. Then Judith argues that a crossing number is the sum of two border numbers: „We have the thirty-one, that, that yes, ... yes that is the same as the eighteen and the thirteen." (396). She represents this at a single, concrete case and adds a general argument for the fact that each border number is used once when one crosses something out, i.e. calculates a circled number with the crossing algorithm: „Well, um, and, and that from the plus numbers, thus from the border numbers that yes, um, eh, ... now each border number, that there comes once into the game, um, if one crosses something out, or so." (396). She emphasizes the general aspect of her argumentation by means of giving a partial justification for the crossing algorithm: „..., and that's why one is allowed in one row to um, now then ... yes then circle only one number. Yes, because otherwise, one number would be contained twice." (396).

Judith's central statements can be structured in the following chain of signifiers (Fig. 53a & 53b)):

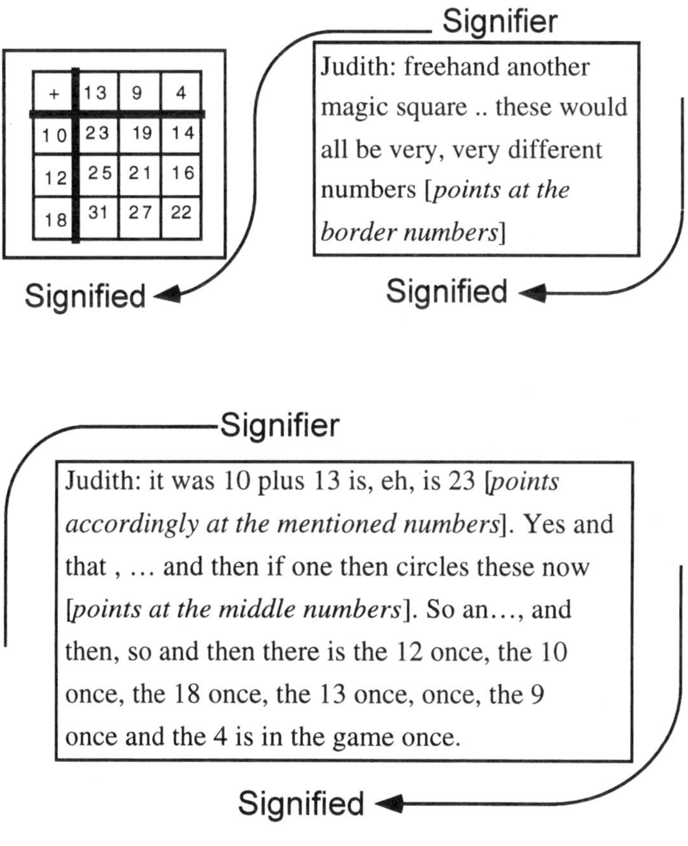

Figure 53a. Judith's chain of signifiers (1. part).

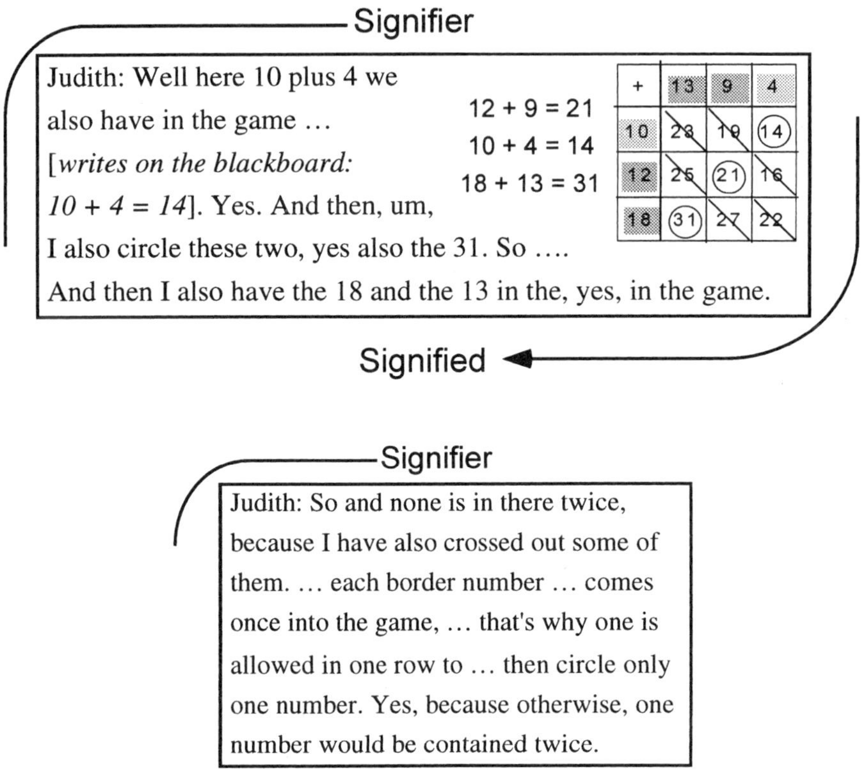

Figure 53b. Judith's chain of signifiers (2. part).

In her first statement, Judith refers to the concrete cross-out number square with border. With the signifier „draw another magic square onto the blackboard“ she intends drawing a „general“ magic square which is not bound to concrete numbers. The intended „general“ cross-out number square can emerge if the border numbers are „variable“ (or are variables). With the signifier „very, very different border numbers“, another constructive form for a „general“ cross-out number square is intended.

Then Judith seems to repeat something concrete and already familiar with the signifier „it was 10 + 13 = 23“. Yet in the context of the generalization („very, very different numbers“) it is most likely meant to communicate information other than merely the calculation itself. This intention is confirmed with the following signifier. In the statement „if one then circles these now [*points at the middle numbers*].“, among other things, the preceding addition exercise 10 + 13 = 23 is interpreted in a new way as a connection between two border numbers and one circled number. Thus it was not the calculation with concrete numbers that was meant, but an arithmetical relation. And in the statement „the <u>twelve</u> once, the <u>ten</u> once, the <u>eighteen</u> once, the <u>thirteen</u> once, once, the <u>nine</u> once andthe <u>four</u> is in the game once“, this intention is transferred to three circled numbers (three circled numbers intentionally mean the magic number). The fact that each border number appears exactly once with the additive composition of the three circled numbers is intended in the signifier „is once in the game".

In the signifier „very, very different numbers", Judith intends a general cross-out number square which can be constructed out of arbitrary border numbers. Referring to the additive construction procedure for the numbers in the square from the border numbers in the signifier „each border number is in the game once with three circled numbers", Judith wants to express the idea that all border numbers are contained once in three circled numbers. Eventually, this means that the sum of the border numbers (which is known to be constant) is equal to the sum of three circled numbers (the magic number). Thus this verbal description essentially represents a complete argument for the problem.

This verbal description is concretized and then symbolized by means of circling and crossing out numbers. Furthermore, the border numbers which belong together and the respective circled number are marked with the same color and also written next to the cross-out number square as an exercise in addition. In the development of the symbolic representation which is accompanied by the verbal descriptions, Judith employs the given, concrete numbers. However, she expresses her intention of generalization by saying that it could also be a very different square, and by means of the repeatedly used statement „... a number is in the game". She uses the concrete situation to emphasize arithmetical relations and connections between border numbers and circled numbers in the cross-out number square; she is not merely engaged in calculating simple addition exercises.

In the contributions 364 and 396, Judith gives hints which are supposed to illustrate that border numbers cannot appear more than once as terms of a sum with circled numbers. In 364, Judith says: „... none is in there twice, because I have also crossed out some of them ..." and means that no border number appears twice as a term of a sum with this procedure. In 396, Judith first repeats the whole argument in her own words and in an abbreviated way. Then she adds a remark that each border number appears exactly once as a term of a sum with circled numbers: „... each border number, that there comes once into the game, um, if one crosses something out, or so. That, that yes, and that's why one is allowed in one row to um, now then ... yes then circle only one number. Yes, because otherwise, one number would be contained twice." By means of this open, general designation she refers to the structure of an unequivocal relation between two border numbers and one circled number.

In the signifier „that's why one is allowed in one row to ... circle only one number. ... otherwise, one number would be contained twice.", Judith formulates an additional, indirect argument for the question why „each border number is once in the game". This complex, open statement can be analyzed as follows. The intention in the remark, „Yes, because otherwise, one number would be contained twice", means to imagine that the relation „circling two numbers in one row means that two equal border numbers are contained" is a possibility and to negate it at the same time as not admissible. In the remark, „that's why one is allowed in one row to ... circle only one number", this negation implies the intended interpretation: „With a circled number in a row, the corresponding border number is used only once with the circled numbers".

In this way, Judith produces a chain of signifiers from the initial signified of the calculated cross-out number square with given numbers. In the first signifier (Fig. 53), she emphasizes the generality of the concrete cross-out number square if one were to take different numbers for the border numbers. In the second signifier (Fig. 53), she uses the concrete numbers as examples for arithmetical relations in the cross-out number square. The third signifier represents a verbal and written down symbolization of the relations between the border numbers and the particular concrete circled numbers. The fourth signifier is a supplement, in a verbal and ostensive way, aimed at illustrating the fact that after circling and then crossing out the other numbers, e.g. in the same row, the respective border number can no longer appear as a term of a sum or that one is only allowed to circle one number per row. In this chain of signifiers, with her descriptions and particular representations, Judith develops a comprehensive constructive argument why the magic number is always 66 or constant.

From an epistemological point of view, with her contributions, Judith constructs a mathematical knowledge relation for cross-out number squares. In her constructions, it becomes clear that the mathematical signs and reference contexts she uses are interpreted in a relational way. The new knowledge relation is worked out proceeding from the construction rule for cross-out number squares as a tabular addition of (here 2 times 3) border numbers. The simple addition rule is specified for the circled numbers and interpreted in a generalizing way: A circled number is made of two border numbers, and each border number appears only once. Thus the idea of an „inversion" of a calculation exercise can be observed again.

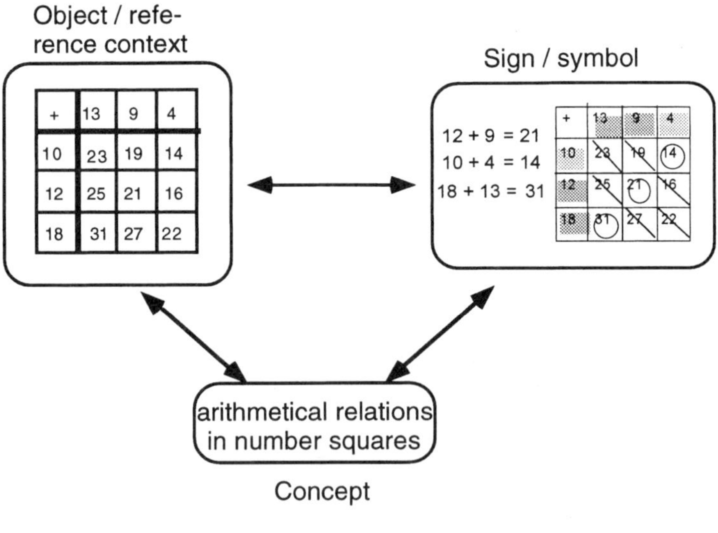

Figure 54. The epistemological triangle: The carrying out of the crossing algorithm.

Addition of two border numbers at a time not only results in the numbers in the cross-out number square, and thus also in the circled numbers, but, inversely, one can also interpret a circled number as containing exactly two of the border numbers.

The construction rule is thus not only an exercise in calculation, but is understood here as a particular arithmetical relation for uncovering connections and specific characteristics in the cross-out number square.

In the diagram of the epistemological triangle (Fig. 54), the essential relations between the mathematical signs / symbols and the reference contexts produced by Judith can be illustrated in the following way. Based on her verbal and deictic interpretation, she develops a complex sign.

Referring to a concrete cross-out number square, Judith has constructed important relations and structures of a general cross-out number square by means of showing and describing. She obtained the generality or the intended general mathematical relation by referring to structural relations through a „negation“ of the concrete aspects („very, very different numbers“) and through a „changed use“ of concrete mathematical connections („10 + 13 = 23“). In this epistemological triangle, the construction of the „sign / symbol“ which emerges by means of the application of the crossing algorithm to the concrete „object / reference context“ of the given cross-out number square is modeled. In a further epistemological triangle (Fig. 55) one can emphasize the general structure of the new sign in a more explicit way.

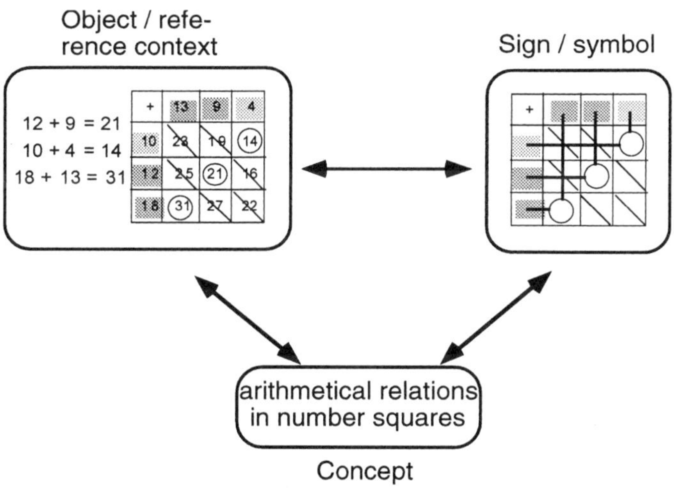

Figure 55. The epistemological triangle: The crossing procedure as a symbolic structure.

It is obvious that Judith does not write down this diagrammatic structure of the sign like this. Yet one can take it from her generalizing statements (... very different numbers, the number is in the game, etc.) that she is not talking about concrete numbers and calculations, but about arithmetical relations which are embodied in her concrete descriptions. The verbal descriptions and the pointing at positions in the square support this structural view of the complex mathematical sign.

Now Judith formulates an additional, indirect argument. She compares the situation in which two numbers in the same row were circled with the situation in which only one number per row is circled. In the epistemological triangle (Fig. 56) one can represent this interpretation as follows (these versions of the cross-out number squares are not employed explicitly, but have only been designated in a general, verbal way).

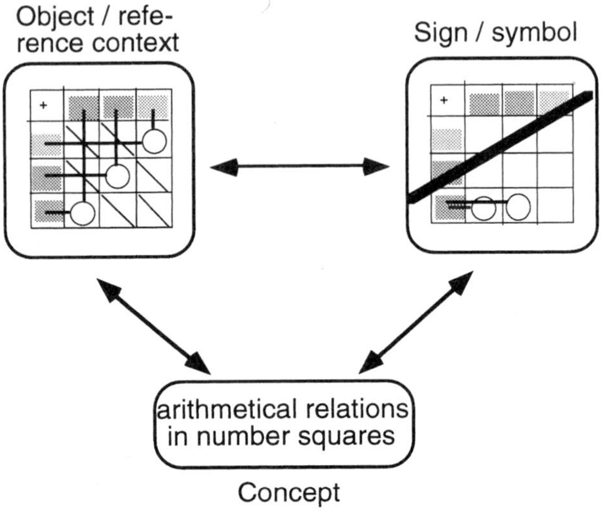

Figure 56. The epistemological triangle: A border number cannot be employed twice.

In the cross-out number square in which two numbers in the same row were circled, Judith constructs a new sign and relates it to a cross-out number square with circled numbers. This sign is of a general kind. The relation between „sign / symbol" and „object / reference context" is exchangeable in the sense that both sides can be seen reciprocally as unfamiliar signs or as explaining reference contexts. The possible new sign is negated, and thus the justification that each border number is only used *once* is strengthened. If one border number in a row was used twice, two numbers in the row would be circled, which is inadmissible.

To sum up the analysis of Judith's knowledge construction, it can be stated that she produces, step by step, the components of an actual mathematical sign / symbol. The new mathematical signs as well as the reference contexts used are interpreted structurally and in a way that is full of relations, and not as concrete numbers or as visible features. Furthermore, the difficult mediation between the signs / symbols and the reference contexts is controlled by means of the conceptual relation of the construction rule for cross-out number squares and the crossing algorithm, and these mathematical relations are further developed and generalized at the same time.

3.1.3 Analysis of the Third Scene from the Episode „Why is the magic number 66 always obtained?"

In this scene, a further constructive justification for the problem „Why 66 is always obtained?" is formulated by the student Kim. During the phase of individual work (see 3.3.1.1) she formulated the idea of wanting to produce the nine numbers in the square by means of dissecting the magic number 66 several times into three numbers. Now she improves her knowledge construction with regard to the realizability of her suggestion.

408 K: Um, mm, one must simply take the sixty-six apart into six numbers. And then one must, if that was cut apart now, still simply put the numbers there somehow, and then calculate, mm, the numbers. And, and then, then one must only still do that with the number square. And then one has the sixty-six as a result everywhere.

409 T: Great, well done! Could you show, please, where you would have to put the numbers which you have cut out.

410 K: [*goes to the board*] Yes simply three of them up there, and then there still a plus [*points at the corresponding positions at the margin of the table*]. And then one must calculate somehow, there for instance nine and ten are nineteen. And then, and if one did that, then the result would always be sixty-six.

Kim expresses an argument by means of constructing „any" fitting cross-out number square proceeding from the magic number 66. To do this, she introduces the dissection of the 66 into six arbitrary terms of a sum (408). She suggests to „cut" the 66 into six terms of a sum, to arrange these arbitrarily as border numbers („simply put [them] somehow") and then to calculate the numbers in the cross-out number square through addition („then calculate the numbers").

Kim makes this explanation clear concretely: She says that one can put three of the six numbers into the „upper" border and the others into the left border (410). For the explanation of the calculation of the numbers within the cross-out number square, Kim refers to the given numbers once again: „And then one must calculate somehow, there for instance nine and ten are nineteen" (410). As above, she briefly adds: „And then, and if one did that, then the result would always be sixty-six." (410). Here, it is not explained once again that all border numbers appear exactly once in the circled numbers. The teacher praises Kim and thus confirms her justification.

One the one hand, Kim formulates a construction procedure of a cross-out number square proceeding from a given crossing sum in a general way. On the other hand, she concretizes this procedure with the numbers in the given square. Her statements can be represented by a sequence of signifiers as presented in Figure 57.

The first signifier formulates in a general way how Kim wants to construct the number square proceeding from the magic number 66. In her own words, she describes how 66 is to be dissected into 6 numbers: she uses the expressions „take apart" and „cut out" for this (cf. her description of this dissection of the magic number into several times three numbers as „divided by 3" in section 3.3.1.1). Then she describes in an non-specific way that these six numbers must be „put somehow" (later it will be clear that these six numbers become the border numbers of a table). Furthermore, the numbers are to be calculated, which probably means the construction of the cross-out number square by adding the border numbers. Thus the construction is described in general terms in the first signifier.

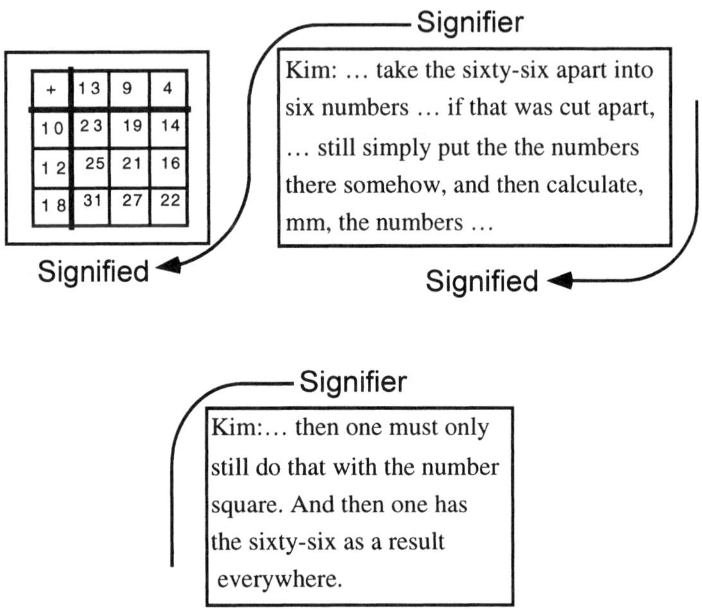

Figure 57. Kim's generalizing signifiers for the construction of a square with the magic number 66.

The second signifier states in a general way that one must „do that with the number square" (obviously apply the crossing algorithm) and then one would have „the sixty-six as a result everywhere" (where this statement is a confirmation of the fact that Kim indeed meant the application of the crossing algorithm with the general remark made before).

This construction description and justification is repeated by Kim at the blackboard with the concrete numbers in the square.

In a mixture of general and concrete references to the given number square, Kim explains her construction with the first signifier (Fig. 58). She points at the margin where three of the six dissected numbers of the magic number 66 are to be written down. Using a sample calculation, „nine and ten are nineteen", with given numbers she refers to the fact that the number square must be calculated from the border numbers. With the second signifier (Fig. 58), Kim generally and briefly refers to the fact that if the calculation of the magic number is applied, the result will always be 66.

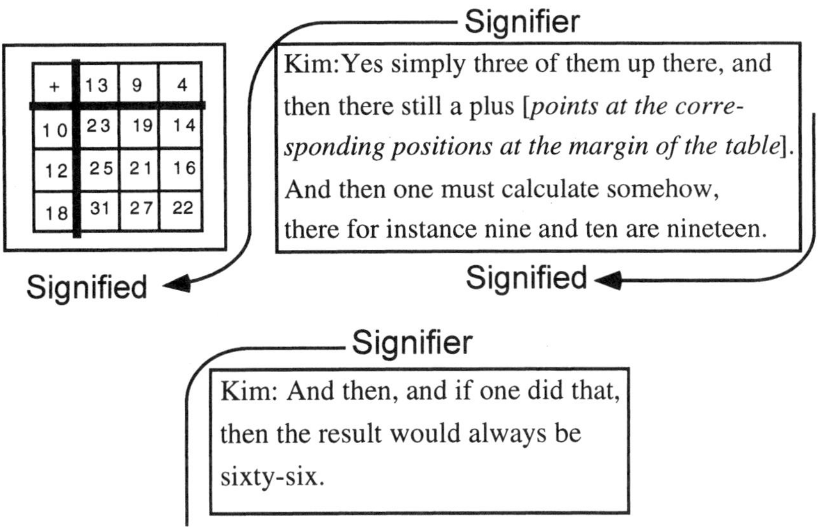

Figure 58. Kim's concrete signifiers for the construction of a square with the magic number 66.

Kim develops her signifiers to describe the production of cross-out number squares from a given magic number in two ways, such that a constant magic number always results in the number squares constructed in this way. The first description is rather open and general whereas the second one is oriented towards the concrete number square. Both descriptions complete each other in the sense that a generality of the construction as well as a justification with its own words on an example becomes possible for Kim. Moreover, the concrete description which is bound to the given numbers confirms the fact that Kim indeed uses a procedure to determine 6 border numbers from the magic number, which, after the calculation of the actual cross-out number square, always leads to a constant magic number (this step of the justification is not explained by Kim any further). This complementarity of open, general description with its own, characteristic words and the example-bound description of the procedure reciprocally support each other and are a suitable form of the situated communication of a general knowledge construction by means of a particular case.

From an epistemological point of view, Kim's knowledge construction begins with a kind of „inversion“ of the statement „The sum of the border numbers is equal to the magic number“: „A given magic number (66) can be dissected into six (arbitrary) terms of a sum which can be used as border numbers for the construction of a cross-out number square.“ In this way, she gives an additional condition for the construction procedure of a cross-out number square over the choice of the border numbers: If one obtains the border numbers by means of dissecting a number (the given magic number) into six terms of a sum, then one can construct such a cross-out number square with a magic number that was chosen beforehand. In her argument, Kim gives this construction: (1) dissect 66 into six arbitrary numbers; (2) put the 6 numbers onto the margin somehow; (3) „calculate“ the cross-out number square; (4) carry out the crossing algorithm. Then, Kim does not mention again the unequivocal connection between two border numbers and one circled number. She immediately states (5) the initial number 66 is the magic number. Thus her argument is complete. The vagueness in the first, general representation (408) of her argument is partly concretized and confirmed in the repetition (410) of the argument formulated in the example. Thus one can show the following construction of the mathematical object „general cross-out number square out of a given magic number“ in the epistemological triangle (Fig. 59).

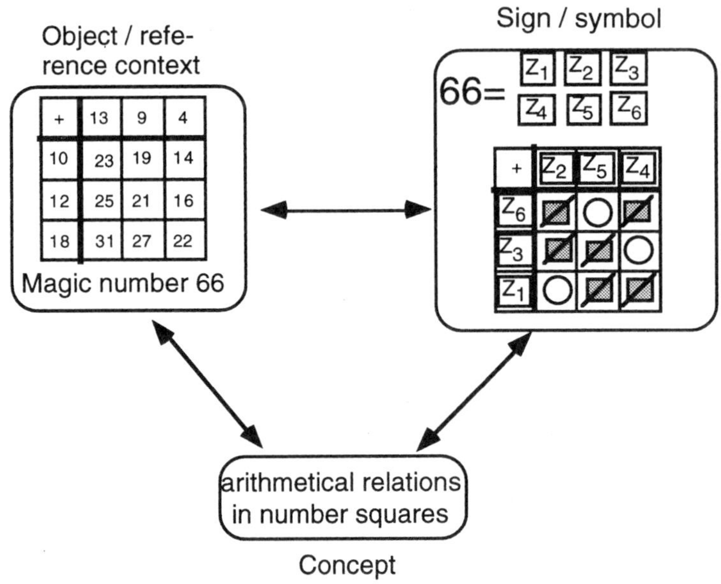

Figure 59. The epistemological triangle: The construction of the border numbers from the magic number.

The representation of the „sign / symbol“ in the form of an abstract visualization which is chosen here has not been constructed in this way by Kim. She has given a verbal description for this and partly confirmed it in the argument (410) by means of pointing at concrete cases in the given cross-out number square. The relation between „object / reference context“ and „sign / symbol“ on the other hand is reciprocally interchangeable. The new relations and structures are interpreted in the given cross-out number square. Reversely, concrete numbers and exercises out of the given reference context are used in order to explain new, general relations by means of examples.

Following up on her earlier argumentation (see section 3.3.1.1) and using her own words, Kim has now described and developed a correct knowledge construction for the production of a magic square proceeding from a given magic number. This construction contains a new knowledge relation for cross-out number squares: The magic number which was dissected into 6 border numbers leads from the addition table to a cross-out number square which has the pre-selected number as its magic number. The analysis made it clear that Kim produced a mathematical relation as a mediation between a mathematical sign and a reference context by means of examples. Signs and reference contexts represent embodiments of structures, and the relation which was developed here has as a prerequisite the mathematical relation of the calculation of the magic number which is now interpreted in a general way (with respect to the relation between the magic number and the different circled number in the number square as first developed by Kim, see section 3.3.1.1).

3.1.4 Analysis of a Scene from the Episode „How can one fill gaps in the cross-out number square?"

The scene which is analyzed in this section comes from the fourth lesson by teacher A. At the beginning of the teaching episode, the teacher presents a 3x3 number square in which a number is missing (Fig. 60).

15	16	17
14	15	16
13	14	

Figure 60. A cross-out number square with a gap.

In relation to this figure, the teacher formulates a problem which is the focus of this episode, but also of the whole fourth lesson of this teaching sequence: What must one do in order to fill a gap (later also several gaps) in a magic square in such a way that the magic square is restored? In order to solve this problem two difficult strategies are developed by children in the course of this episode. On the one hand, possible border numbers for the magic square are constructed in order to determine the missing. In the following scene, Kim makes another suggestion. She wants to use the magic number (the constant sum of three circled numbers) in order to determine the missing number. The knowledge construction carried out by Kim takes place at several points during the lesson; accordingly, only her essential contributions are quoted.

During the phase of the reconstruction of border numbers to determine the number in the gap, Kim makes the following statement.

60	K	Eh, one could also do it like this, that one, one would do that now already. Then and then [one] doesn't have the number yet. And then [one] calculates together what's missing there. And then one can also calculate how, what belongs there.
61	T	What do you mean by that?
62	K	Ehm, that one now circles the thirteen, for example. And then would just cross out the fourteen. ...
63	T	So you ha...

64 K [*incomprehensible*] ... fifteen. And then [one] circles the fiftee..., no, yes, the fifteen. [One] circles the fourteen and the sixteen. And then [one] circles the seventeen and crosses out the sixteen.

65 T Listen, Tim, eh, Kim! That‘s a really neat trick! We‘ll get to speaking about that directly.

Kim makes her proposal in a compressed, incomprehensible form: one calculates something together, which is missing there (the magic number??); and then one can calculate what comes in the empty field (60). Kim is about to explain it. By concretizing her suggestion with the help of the given numbers (64) it becomes clear that she first considers the calculation of the magic number of the incomplete square. She wants to circle the three numbers in the diagonal: „... that one now circles the thirteen, for example. And then would just cross out the fourteen. ... fifteen. And then [one] circles the fiftee..., no, yes, the fifteen. [One] circles the fourteen and the sixteen. And then [one] circles the seventeen and crosses out the sixteen.“ (62, 64). The teacher praises this suggestion; yet she wants to pursue the strategy of the border numbers first and then come back to this proposal.

Before Kim begins her actual explanation, the calculation of the magic number from the three numbers 13, 15 and 17 (cf. Fig. 61) has been thoroughly carried out. The crossing exercise was noted and calculated: 13 + 15 + 17 = 45.

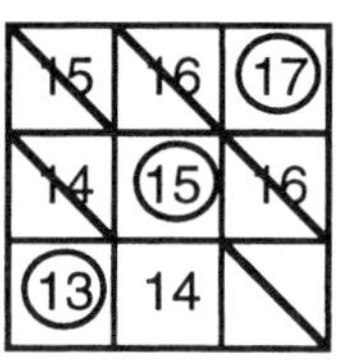

13 + 15 + 17 = 45

Figure 61. The calculation of the magic number in a cross-out number square with a gap.

Here Kim begins with the actual description of the calculation of the missing number.

147 K: And then one could already do it this way. One circles the fifteen [*points at the fifteen in the first line*] and this fifteen [*points at the fifteen in the second line*] Then [one] adds them. And then one still calculates how much there must be up to forty-five.

Kim explains how one can determine the missing number with the help of the magic number which was determined before: „And then one could already do it this way. One circles the fifteen [*points at the fifteen in the first line*] and this fifteen [*points at the fifteen in the second line*]“ (147). She chooses another combination of crossing numbers or of „positions“ for crossing numbers, namely the numbers on the side-diagonal. „... Then [one] adds them. And then one still calculates how much there must be up to forty-five." (147). The sum of these two numbers does not yet result in the magic number; one must still complete it up to 45.

As Kim's explanations remain incomprehensible for several students, the teacher asks her to explain her suggestion in more detail. In order for her to do so, the teacher has restored the initial form of the number square, i.e. she has wiped out the changes made by Kim by circling and crossing out.

161 K: First one calculates, one first calculates these numbers, that I have, which are there, what is their result. And then..., and then one calculate

162 T: So. Now Kim explains how it goes on! Kim.

165 K: These three, oh, yes, this, this and then afterwards one calculates fifteen [*circles the 15 in the second line*], one takes this way. [One] crosses out that, and that. And [one] crosses out that, and that. [*crosses out numbers which are in the same line or column as the 15*] Then one still takes the fifteen. [*circles the 15 in the first line*] [One] crosses out the seventeen and the thirteen. [*crosses out the numbers in the same line or column as the 15 that haven't been crossed out yet*] And then one circles this here, this here. [*circles the empty field*] And then one has to calculate here, fifteen and fifteen make thirty; how much is left up to forty-five?

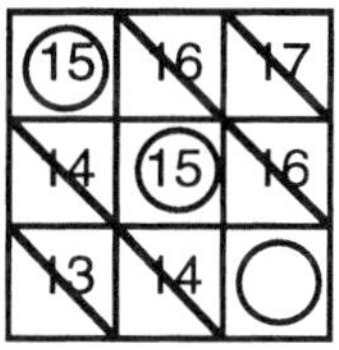

13 + 15 + 17 = 45

Figure 62. The magic number in a cross-out number square with a gap.

166 T: Just write that down like this as an addition exercise! As you just did it, with a gap if you want to. [*Kim wants to wipe out the exercise „13 + 15 + 17 = 45"*] No, just leave that below!
With the support of the teacher and of other children, Kim writes down the following arithmetical equation.

176 K: [*adds the equals sign to the exercise „15 + 15 + __ = „*] Yes. Fifteen and fifteen are already thirty. [*points to the two terms of the sum „15"*] And then [one] must, one must only calculate yet, ehm, how much up to forty-five

177 T: Write the forty-five down already! … [*Kim writes down the sum „45" to the exercise, 9 sec break*]

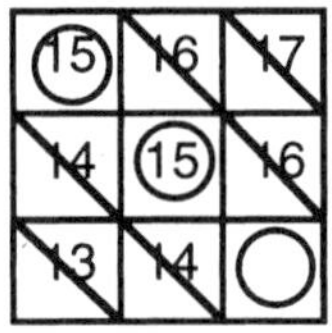

13 + 15 + 17 = 45

15 + 15 + __ = 45

Figure 63. An arithmetical equation for calculating the number in the gap.

At the beginning (161) Kim repeats the necessity of the calculation of the magic number in a verbal, partly unclear way: „… one first calculates these numbers, that I have, which are there, what is their result.“ Then she begins concretely with the application of the crossing algorithm to the numbers 15 (in the second row) and 15 (in the first row). Through this procedure, a visible system of mathematical signs which corresponds to the familiar, visual scheme of the algorithm for the magic number is gradually developed in the reference context by means of the graphical signs of circling and crossing out.

In the statement „And then one circles this here, this here.“, the crossing algorithm is intentionally applied to an (unknown) third number, namely the missing number in the empty field. One intention of the second remark, „then one has to calculate fifteen and fifteen. This makes thirty; how much is left up to forty-five?“, is the calculation of the magic number from the three circled numbers: 15 and 15 are 30. But one cannot continue calculating with the third circled position. Thus, using an „inverse“ method, the unknown position is calculated in such a way that the number missing from 30 up to 45 (the magic number) comes in this position. The „scheme“ of the calculation of the magic number is transferred to the empty spot and thus a new sign is intended and also represented iconically. With this „sign“, the magic number cannot be calculated in the familiar way. Furthermore the magic number is already known. Thus the calculation in the scheme is interpreted in another way: The number which is missing up to the magic number comes in the empty field.

With the help of the teacher, Kim writes down the addition exercise to determine the missing number: „15 + 15 + __ = 45“. Thus it is intentionally referred to the crossing algorithm for three numbers and at the same time to the „unknown“ number in the empty field. In this way a new (open) sign is created and written down using a mathematical notation.

The knowledge construction carried out by Kim at several points during the course of the lesson can first be represented as the following chains of signifiers (Fig. 64).

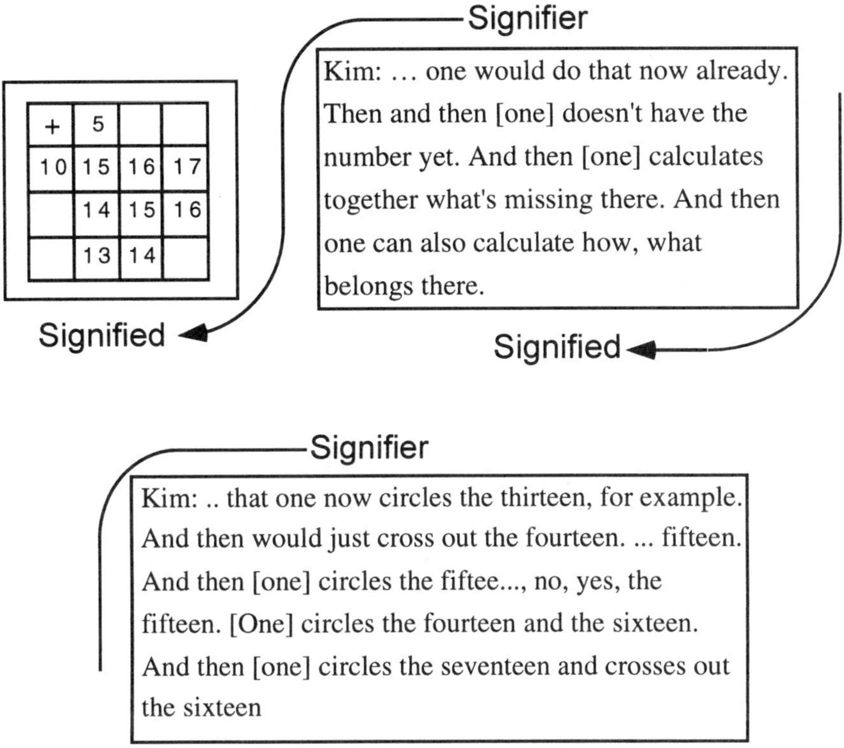

Figure 64. Kim's general signifiers to determine the magic number.

From Kim's first signifier, one can only take the very general information that something is to happen to the number square. Perhaps it can still be reconstructed so that the magic number could be found in this way. However, no general procedure to determine the magic number can be recognized in this description so far. The statements made by Kim remain too open and too ambiguous.

The second signifier formulated by using the given numbers allows for several possible interpreta tions of the first, general signifier. With this, Kim intends to apply the crossing algorithm to the three numbers on the diagonal: 13, 15 and 17. Thus the first step of her justification becomes visible. Furthermore the intention of finding out the missing number by means of the crossing sum (the addition exercise 13 + 15 + 17) becomes apparent.

Using the second signifier, one can carry out the following explanation of the first interpretation. With the first part of the signifier „Eh, one could also do it like this, that one, one would do that now already. Then and then [one] doesn't have the number yet.", Kim could have intended in a very general way: „One can already carry out the crossing algorithm when one has not yet found the number in the gap."

With the second part of the signifier „And then [one] calculates together what's missing there. And then one can also calculate how, what belongs there" could again be intended in a very general way: „Add the three numbers and calculate what is missing in the gap. Like this (with the addition exercise for the magic number), one can calculate what belongs in the gap".

This possible explanation of the first, very general and open description of a procedure to determine the missing number finds a confirmation in the concretization of the suggestion by Kim as she carries it out at a later point of the lesson. Before Kim begins with her actual construction description. The crossing algorithm was applied to the three numbers 13, 15 and 17 in a detailed way and then the crossing exercise was written down and calculated: 13 + 15 + 17 = 45. At this point the essential description to determine the magic number, which is represented in the following signifier, takes place.

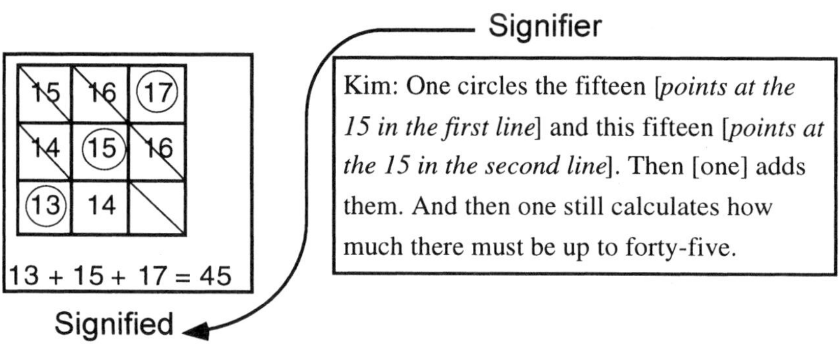

Figure 65. Kim's signifier to determine the missing number.

The intention of the signifier „One circles the fifteen and this fifteen. Then [one] adds them", is to apply the crossing algorithm to two numbers on the diagonal. The second signifier „And then [one] still calculates how much there must be up to forty-five", can be interpreted in the following way. From the sum of 15 and 15, one must still calculate up to 45 (determine the difference) a number, which then belongs in the empty field. Usually, one must apply the crossing algorithm to three numbers: the two 15s and the empty field. However, in this case, the number is not there, but the result, the magic number, is already known, 45. One can say that in her description Kim formulates the following statement productively in a new way. The calculation „And [one] then adds the number in the empty field in order to obtain the magic number 45", which is necessary for the correct application of the crossing algorithm, but not possible as a normal calculation, is formulated by Kim in this workable way of proceeding: „And then [one] still calculates how much there must be up to forty-five".

Kim needs to further explain her suggestion. To do so, she directly applies the algorithm of circling and crossing out. The teacher has restored the original form of the square for this. In a chain of signifiers, one can represent Kim's statements as follows (Fig. 66).

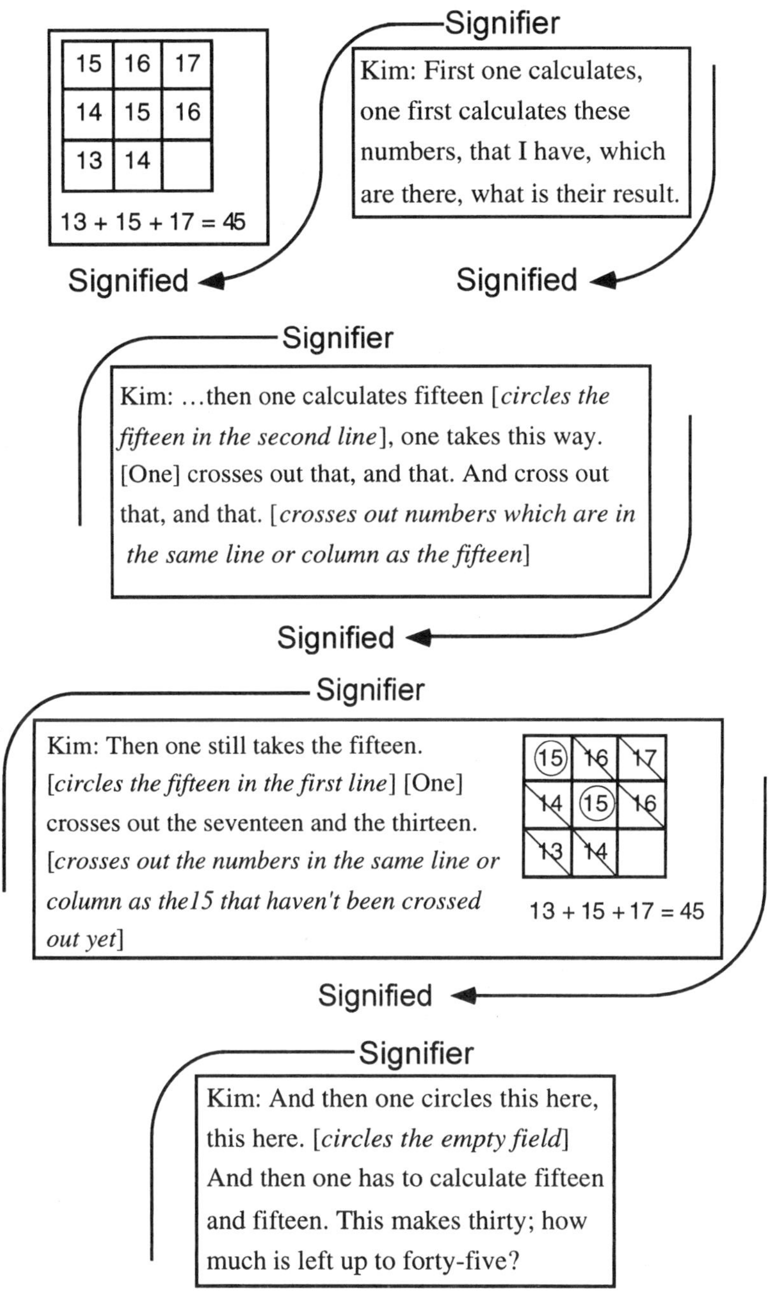

Figure 66. Kim's chain of signifiers to determine the missing number.

First Kim again communicates a signifier: „...one first calculates these numbers, that I have, which are there, what is their result." This seems intended to say that the calculation of the magic number must be carried out first, for instance as it has already been done in the addition exercise. Then Kim begins with the concrete application of the crossing algorithm to the numbers 15 (in the second line) and 15 (in the first line), as it is presented in the second and third signifier step by step. By means of the circling and crossing out a graphical change of the number square takes place, thus a graphical mathematical sign develops.

In the fourth signifier „And then one circles this here, this here", the crossing algorithm is intentionally applied to an (unknown) third number, namely the missing number in the empty field. In the statement „then one has to calculate fifteen and fifteen. This makes thirty; how much is left up to forty-five?", the calculation of the magic number out of the three circled numbers is intended on the one hand: 15 and 15 is 30. But one cannot continue calculating with the third circled position. On the other hand, it is now calculated in an „inverse" way with the unknown position in such a way that the number missing from 30 up to 45 (the magic number) comes into this position. The „scheme" of the calculation of the magic number is transferred to the empty place and thus a new sign is intended and also represented iconically. With this mathematical „sign", the magic number cannot be calculated in the familiar way; the magic number is already known. Thus the calculation in the scheme is interpreted in another way: The number which is missing up to the magic number comes into the empty field.

Finally, after the teacher's request, Kim writes the equation with the unknown number onto the blackboard (Fig. 67).

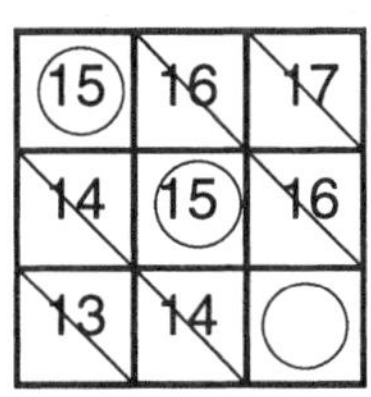

13 + 15 + 17 = 45
15 + 15 + __ = 45

Figure 67. Kim constructs a mathematical sign in order to determine the missing number.

By means of her contributions at different occasions during the course of the whole episode, Kim has presented her own, complete knowledge construction which represents a procedure to determine a missing number in the cross-out number square. From an epistemological perspective, in the whole construction, Kim essentially produces two mathematical signs. Once, she verbally expresses the idea that one can determine the crossing sum in the present incomplete cross-out number square as well. Furthermore, she applies the procedure to determine the crossing sum also to the empty field in the square.

The first sign can be described in the epistemological triangle as follows (see Fig. 68). Here, attention must be paid to the fact that Kim gives the calculation of the magic number only by means of pointing gestures and speaking, but does not carry out an actual action of circling and crossing out. From an epistemological perspective, Kim tries to construct the missing number as a relation between an addition exercise to determine the magic number and another calculation in a way which is still open and not yet completed.

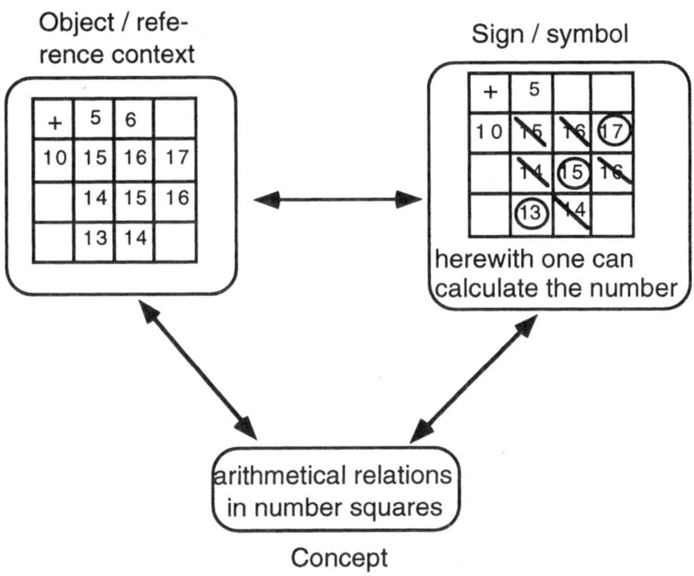

Figure 68. The epistemological triangle: The magic number in the incomplete number square.

The new mathematical relation between the magic number and the missing number in the interrelation between „object / reference context“ and „sign / symbol“ is here only partly hinted at and it is not yet worked out. This construction to determine the magic number in an incomplete square can only be understood as an important component in the following construction to determine the missing number.

From an epistemological point of view, looking at the explanations made by Kim, one can especially state that, in her argumentation, she constructs essentially new knowledge when operationally working with the unknown number in the empty field. One cannot only calculate with concrete numbers, but also transfer the scheme of the crossing algorithm to arbitrary fields with or without numbers. Then one can operationally vary the arithmetical calculation if all numbers are known except for one number remaining unknown. Kim already represents her argumentation in a certain „logical“ sequence. First, the magic number is determined from three possible crossing numbers (the purpose cannot be seen here yet). Then the algorithm is applied to two numbers and the empty field, and in order to do so, one eventually requires the knowledge of the magic number (here the necessary calculation of the magic number becomes obvious): The new constructed mathematical knowledge in Kim's argumentation can be characterized using the epistemological triangle.

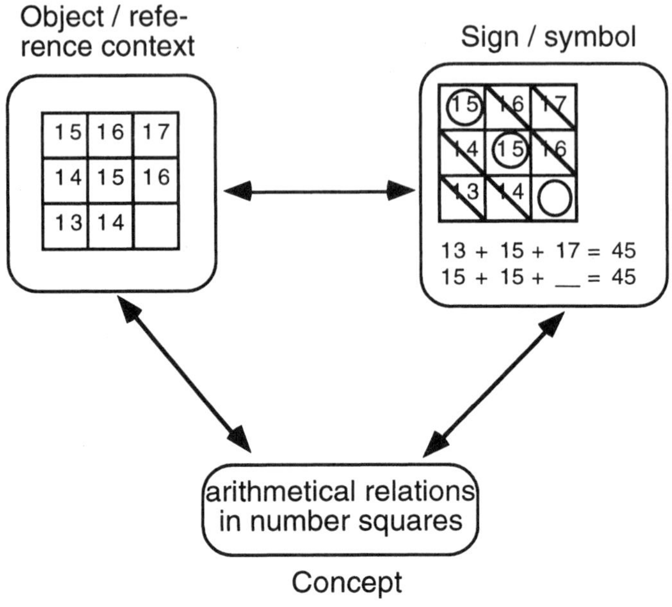

Figure 69. The epistemological triangle: A connection between the missing number and the magic number.

The unknown, missing number or „variable" is symbolized in two ways, once in the cross-out number square as a „circled field" to determine the magic number, and second in the calculation as a „missing number" in an addition exercise in which the magic number to be determined – the result – is already known. In this representational context, one works with a „mathematical unknown" in a situated way. This unknown number is constructed in an arithmetical equation as new knowledge, as a new mathematical relation, and bound into the structural, arithmetical relations of the cross-out number square.

The communicative and epistemological analysis of Kim's knowledge construction makes it clear that the components of an actual mathematical sign / symbol are produced in the successive partial steps. The new mathematical sign and the accompanying reference contexts are used by Kim in a structural way which is full of relations. The mediation between the signs / symbols and the reference contexts is controlled by means of the conceptual relation of the crossing algorithm to determine the magic number and this mathematical relation is at the same time further developed and generalized for the determination of a missing number.

3.2. Teaching Episodes from Teacher B's Instruction

3.2.1 Analysis of a Scene from the Episode „How do the number walls change when the four base blocks are interchanged?"

In this section, I shall look in detail at a classroom scene from the second lesson carried out by teacher B in the learning environment of „number walls". The day before, the teacher made her 4th grade students familiar with the exercise format of the number walls. Among other things, students were calculating missing numbers in a four-layered number wall where the numbers for the base blocks were given. These were 35, 45, 55 and 65, which the teacher, however, wrote down in different sequences for individual students. Then each child had cut his or her number wall into a puzzle, which then had to be correctly assembled by his or her neighbor. In a whole class reflection on this exercise, it had already been mentioned that different goal numbers had resulted from the number walls in spite of the same base numbers, and individual children had already observed that this was related to the different succession of numbers in the base layer of blocks.

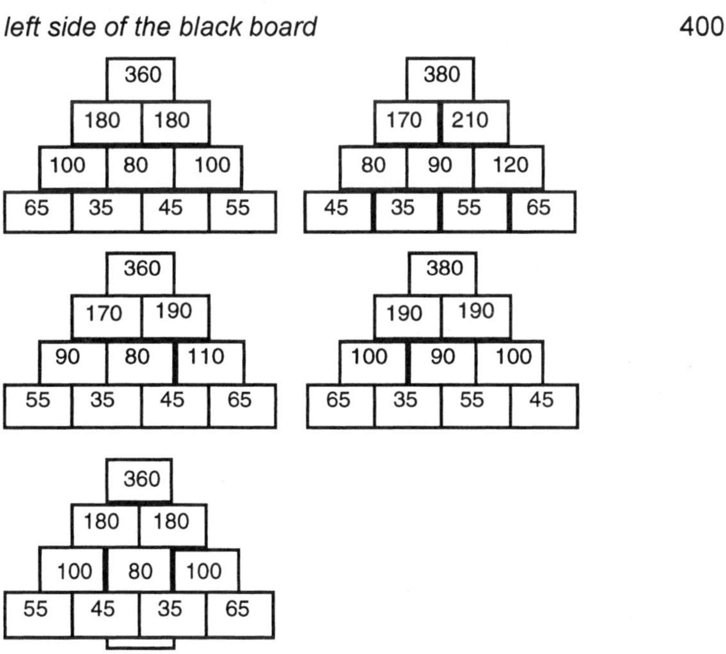

Fig. 70a. Nine number walls with four interchanged base blocks (left side).

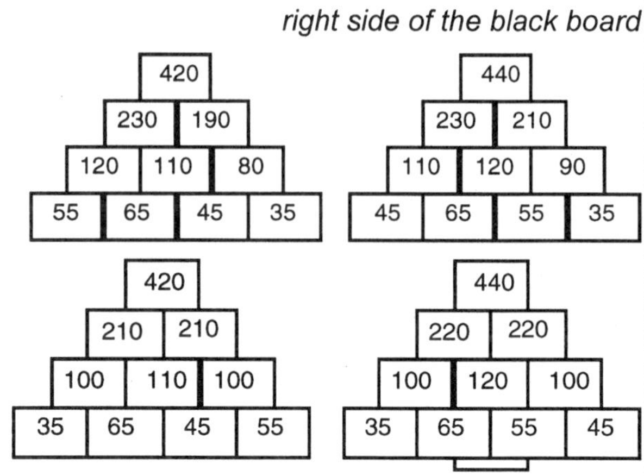

Fig. 70b. Nine number walls with four interchanged base blocks (right side).

After having looked at other number walls during the first 15 minutes of the present second lesson, the teacher comes back to the number walls that were worked on the day before. She asks children to arrange the four given numbers (35, 45, 55 and 65) on the base blocks in such a way that, on the one hand, the smallest and, on the other hand, the biggest goal block is obtained. Attached to the board, there are altogether 9 number walls (see Fig. 70). Children have produced them in pair work. Through – also systematic – interchange of the four base blocks, they were expected to construct number walls with big and small goal blocks.

The topic of the first phase of the discussion, following this pair work, is „Big and small goal numbers in a four-layered number wall". In this scene, Julia and Timo name the first striking features and take an initial approach to their justification.

After several children have hinted at further striking features and made more observations, the teacher asks for a justification of the connection between the magnitude of the middle block in the second row of the wall and the magnitude of the goal block. The student Nele and another child explain the connection in several ways.

Following this scene, the class discusses the question, which of the numbers of the base blocks can be interchanged without changing the number in the goal block. In the discussion with the children, it is pointed out that one can interchange the two border blocks and the two middle blocks without producing a change in the goal block.

In the scene analyzed in more detail in the following, reasons for the change in the goal block when interchanging the base blocks are examined. The teacher refers to the fact that the border numbers in the second row change when the two inner base blocks are interchanged, and she asks why the goal block nevertheless remains the same. With her question, the teacher opens the third episode of this scene (contributions 174–186). In this episode, Timo and Nele try to find an answer to the problem formulated by the teacher and also to justify their answer.

174 T Mh, I'll repeat the question once again, Sasha, since it is very difficult. … Well, when we change in the middle, [*points at the second and third block, „45“ and „35“, subsequently at the sixth block „80“ of the lower number wall with the goal number „360“]* comes here [*points at the second and third block „35“ and „45“ of the middle number wall with the goal number „360“, subsequently she points at the sixth block „80“*] always the same result, sure. You found that out well. But if we change here [*points at the third and fourth block „35“ and „65“ of the lower number wall with the goal number „360“*], this block changes. [*points at the seventh block „100“ of the lower number wall and at the seventh block „110“ of the middle number wall with the goal block „360“*]. Still the goal block is the same. [*8 sec pause*]

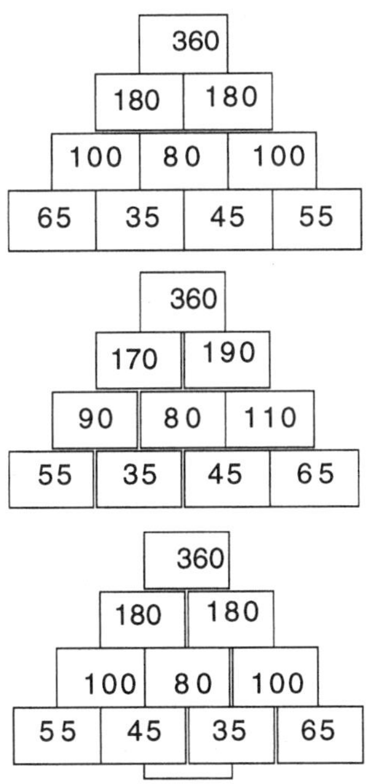

Figure 71. Part of the blackboard image.

174b Sa Because eh…. That's always the same.
174c T Timo!

175	Ti	That's because, [*goes to the black board*] well, yes, ehm here, here is one hundred, one hundred [*points at the fifth and the seventh block „100" of the lower number wall with the goal number „360"*]and here is one hundred and ten and ninety. [*points at the seventh and then at the fifth block in the middle number wall with the goal number „360"*] That compensates each other. There is ten more and there is ten less. Then it is the same again. ... [*he points at the seventh block „110" and the fifth block „90" in the middle number wall with the goal number „360"*]
176	T	Who can explain that once more? Then I can write it down again, Nele!
177	Ne	Well that is always. [*goes to the black board*] Here, the tens are distributed in another way, [*points at the seventh block „110" in the middle number wall with the goal number „360"*] then here ten less [*points at the fifth block „90"*] and there ten more. [*points at the seventh block „110"*] And here it is the same then. [*points alternately at the seventh and the fifth block „100" of the lower number wall with the goal number „360"*]
178	T	[*points first at the fifth block „100" of the lower number wall and then at the fifth block „90" of the middle number with the goal number „360"*] Ten less. [*writes „–10" into the fifth block of the middle number wall with the goal number „360"*] Can you see?
179	S	The difference is the same.
180	T	A little small, isn't it? And here accordingly then? [*points at the seventh block „110" of the middle number wall with the goal number „360"*] ...
181	S	Ten more.
182	S	Ten more.
183	T	Yes. [*writes „+10" into the seventh block*]

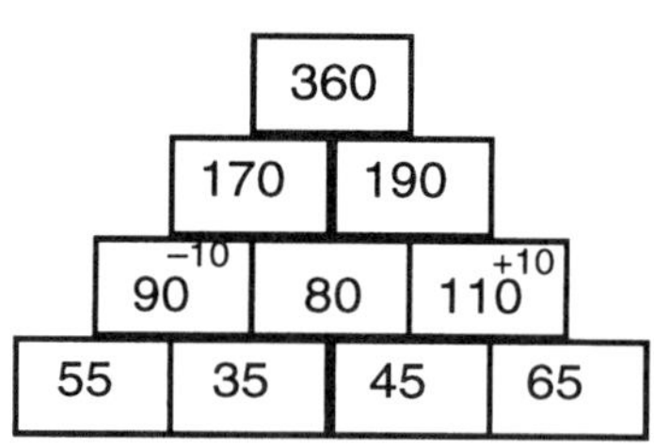

Figure 72. Arithmetical changes in the number wall.

This scene contains, essentially, the knowledge construction of two children. Thus this interaction phase can be structured and summarized as the following sequence.

Phase 1 (174) Repetition of the problem by the teacher

Why doesn't the goal block change when the border blocks in the second row of the wall are changed by means of interchanging the middle base blocks? The teacher points at the 2nd and 3rd block in the lowest and in the middle number wall with the goal numbers „360“: When these are interchanged, then the result – she points at the sixth block „80“ – is always the same. But if the numbers of the 3rd and the 4th block in the lower number wall are interchanged, a change in the 7th block results as compared to the middle number wall. Still – as the teacher says – one obtains the same goal block. Why?

Phase 2.1 (175) Timo compares the corresponding border blocks of two number walls: They compensate each other.

Timo points at the 5th and the 7th block in the lower number wall: here it is 100. Then he points at the 5th and 7th block in the middle number wall as well: 110 and 90. He says that they compensate each other, once, ten more and once, ten less.

Phase 2.2 (176/177) Nele repeats the explanation.

Nele explains this once again. She points at the 7th block „110“ in the middle number, and she says that the tens are distributed in another way: Here 10 more and here (5th block) 10 less. In the lower wall, the 5th and the 7th block are equal.

Phase 2.3 (178–183) The difference of „+10“ and „–10“ is marked

The teacher compares the „lower“ 5th block with the „middle“ 5th block and says: 10 less. She notes „–10“ in the 5th block of the middle wall. Then she points in an inquiring manner at the 7th block „110“ in the middle wall. Students shout „10 more“, the teacher notes „+10“ in the 7th block of the middle wall.

The knowledge constructions in number walls which are formulated by the two students in their statements can be represented with the following signifiers. Seen in a very general way, the blackboard image with the 9 number walls (Fig. 70) represents the signified or the reference context. In this spacious diagram, the teacher as well as the two students concentrate on the following dissection of the three number walls which can be seen on the left. The teacher had formulated the problem with reference to these three walls: „Why don't the goal blocks in the numbers with the goal block 360 change even though different border numbers can result in the second row of the wall?“

The justification of the problem stated by the teacher takes place essentially by means of comparisons of the differences between the corresponding numbers in the walls.

With the first signifier, Timo presents these observations: In the lower number, both the 5th and the 7th block are labeled 100, and in the middle wall, accordingly, 90 and 110. In the second signifier, he then communicates that these numbers compensate each other, once, 10 more, and once, 10 less. Thus the present numbers are implicitly compared to the number „100". The signifiers produced by Timo are of a verbal and deictic kind, as he points at blocks in the number wall and names the numbers which are there at the same time (the pointing is indicated by means of numbered arrows in the diagram, cf. Fig. 73). Timo discusses neither the reasons for this „arithmetical equality" of „10 more" and „10 less" nor why this leads to the same goal number (for instance because it is again about border numbers which are added once). He mentions the striking visible arithmetical features, but does not develop a deeper justification for the invariance of the goal number.

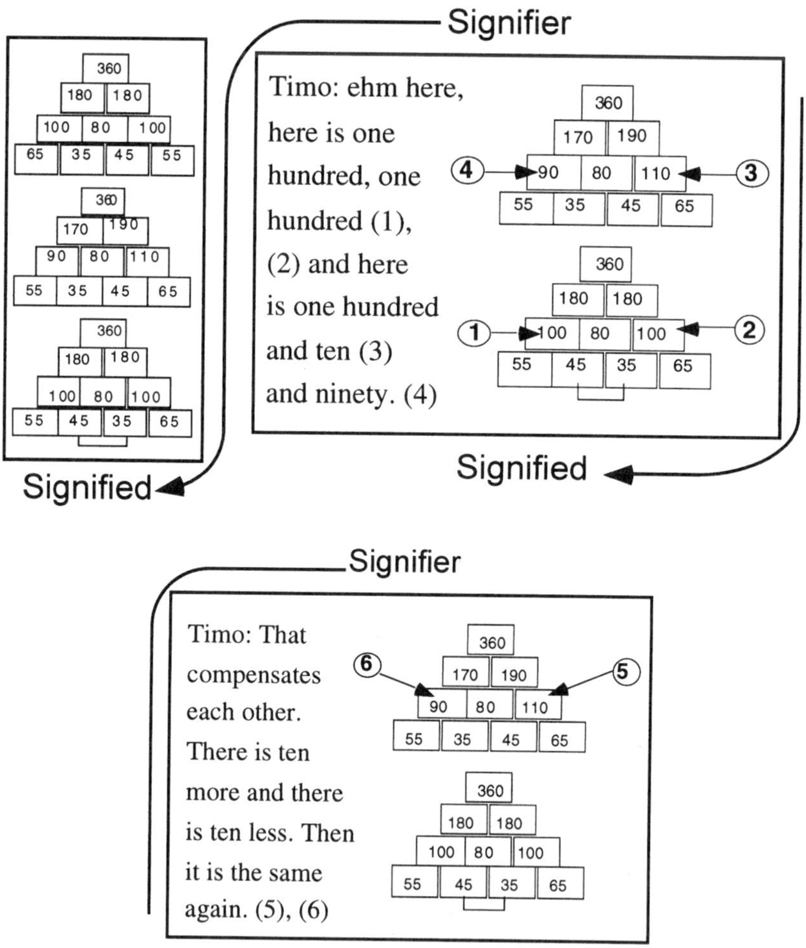

Figure 73. Timo's signifiers to justify the compensation of numbers.

Nele's explanation of the constancy of the goal numbers is similar to Timo's (Fig. 74). She seems to take up the „ten" from Timo's statement and she says and indicates in her first signifier that the tens are distributed in another way. She also states that the 5th block (middle wall) comes to 10 less, and the 7th block to 10 more. The corresponding blocks in the lower wall are equal, she says while pointing at them. This statement represents a repetition of the observation of arithmetical differences and „equality“ in the present number pattern. Like Tim, Nele develops no deeper reasons why the „arithmetical symmetry“ implies the constancy of the goal block.

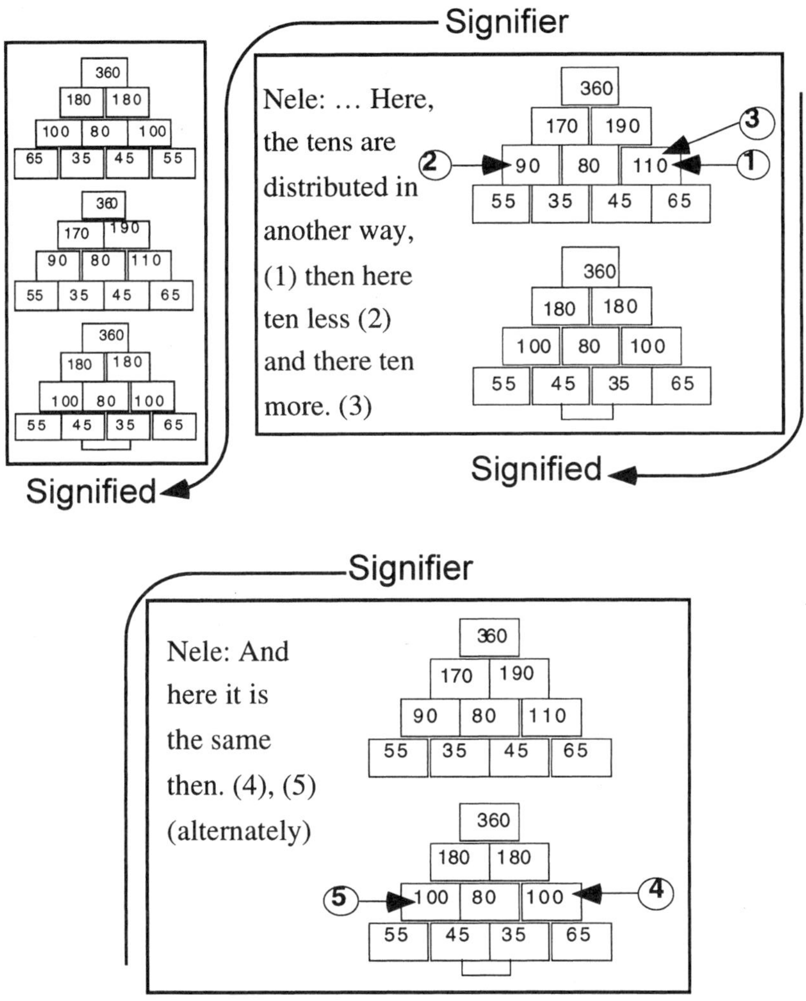

Figure 74. Nele's signifiers to justify the compensation of numbers.

The teacher confirms the „arithmetical symmetry“ which has been named by the two children by means of repeating it and writing down „–10“ as well as „+10“ into the 5th and 7th block (cf. Fig. 72).

Essentially, visible arithmetical features of the concrete numbers in the two number walls are named in the signifiers communicated by the two children; namely the numbers 100, 100, 90 and 110. The signifiers are of a verbal and deictic nature; they refer directly to the signified of the three observed number walls on the board. By means of pointing at comparisons, relations between the numbers are emphasized as important in an intentional way. These relations and structures are essential for mathematical knowledge. However, it is doubtful whether these relations between certain numbers, which are shown by both students by way of concrete numbers, are connected to the construction relations of number walls, thus whether these students indeed talk about actual *mathematical* relations and about *mathematical* signs / symbols here.

From an epistemological perspective, by means of speaking and pointing Timo and Nele construct new signs which are produced as relations between numbers in number walls. These two signs by Timo and Nele are of a similar kind; their representation in the epistemological triangle resembles the representation in the communicative model.

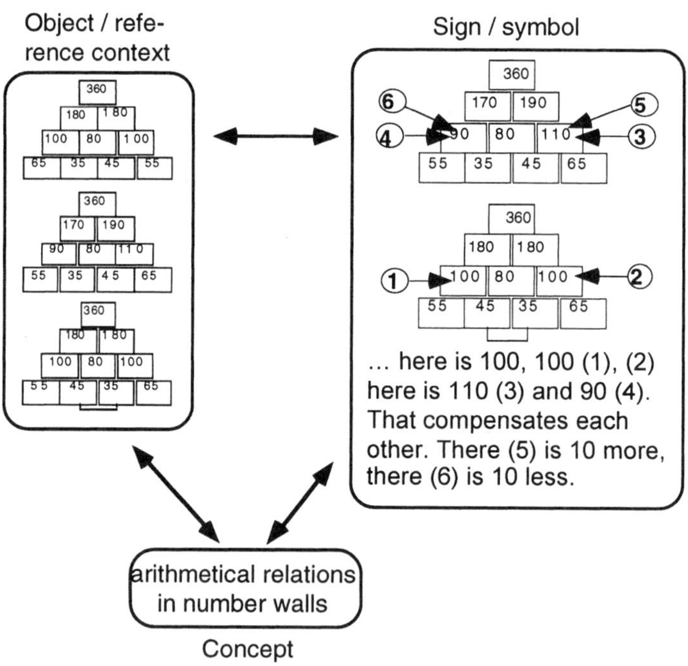

Figure 75. The epistemological triangle: Timo's construction of a compensation relation between numbers in different number walls.

The compensation relation produced by Nele in a similar way results in a comparable epistemological triangle for this example of classroom interaction.

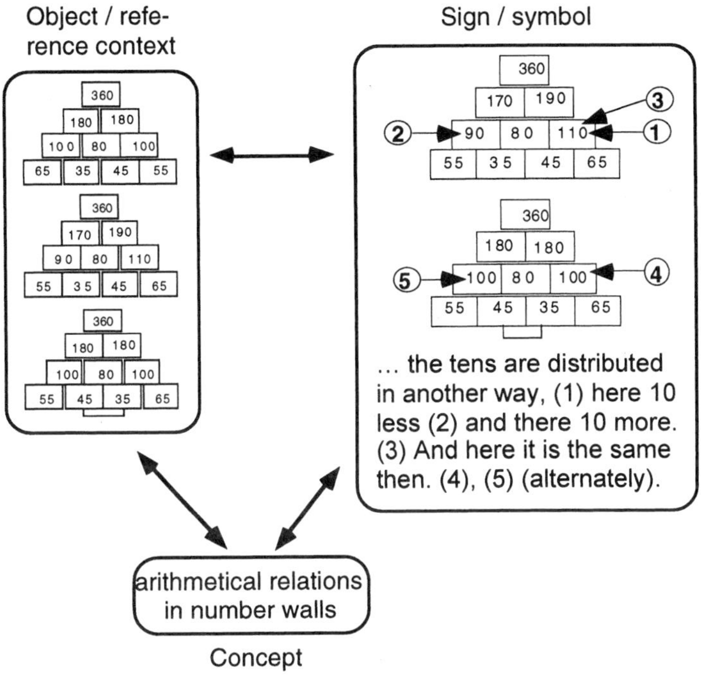

Figure 76. The epistemological triangle: Nele's construction of a compensation relation between numbers in different number walls.

From an epistemological perspective, the knowledge relations which have been produced in this short classroom scene can be discussed in the following way. The teacher begins with an open – partly incorrect – problem: „When the two middle blocks in the base row are interchanged, then the middle block in the second row does not change; when the 3rd and the 4th block (of the basis) are interchanged, then the 7th block (the right border block of the second row of the wall) changes. Why?". The change in the 7th block, which is hinted at here, is not the result of an interchange of the 3rd and 4th block (then the 7th block above them would remain unchanged), but results also from the interchange of the two lower middle blocks. Still the children react to this question in a probably expected way in the interaction.

Timo constructs a possible, justified relation by means of using a striking arithmetical feature: He compares the 5th and the 7th block, states that the two blocks „compensate" each other, once, 10 more and once 10 less. He says that „... that's because", i.e. that is why the goal block remains unchanged. This relation is also used by Nele in order to justify the invariance of the goal block: On the two blocks in question, the tens are distributed in another way, and thus they are the same (together), and thus the goal block has not changed. The teacher confirms this argumentation by means of writing down „–10" and „+10" in the 5th and the 7th block. Thus – one could assume – the justification follows that the goal blocks remain unchanged.

The interactively developed argument can be summarized as follows: The goal block remains unchanged because – in spite of the change – the border blocks in the second row „compensate“ for each other. This argument represents only a possible partial aspect of a complete argumentation. Firstly, the „compensation“ has only been observed and stated. How does this compensation occur and what justification could one give for this? Further: Why does this compensation lead to the invariance of the goal block? Obviously, the lower number wall with the goal block „380“ (cf. Fig. 70) also has this „compensation“ at the 5th and 7th block with 100 respectively as compared to the number wall which was talked about, yet it has a different goal block. Thus other conditions must be added. Can it also be justified that the goal block remains unchanged with such a „compensation“ between the 5th and the 6th block?

The arguments in this interaction refer to the arithmetical structure of the numbers in the wall. An arithmetical compensation is used from which the invariance of the goal block is inferred. This argumentation does not pay enough attention to the structural relations between the blocks or numbers in a four-layered number wall. The argument could be completed only by means of a deeper observation of the underlying relations when constructing the wall.

The communicational and epistemological analyses of Timo‘s and Nele‘s knowledge construction show that only one condition for the construction of a mathematical sign is fulfilled in the mediation between the signs / symbols and the reference contexts. In the two suggested constructions, it is taken into consideration that on the side of the signs / symbols and the reference contexts, it is about relations and structures and not about a mediation between directly visible features. The mediation constructed by the two students this way, however, does not refer to conceptual construction conditions for number walls, but to general, arithmetical compensation relations between the concrete, observed numbers. Thus, no conceptual knowledge which is relevant for the problem situation in question is used in an emerging way or further developed in the construction of new mathematical signs. In both cases, no true mathematical sign / symbol for the present problem is developed; the suggested signs are not connected with the present conceptual knowledge.

3.2.2 Analysis of a Scene from the Episode „How does the goal block change when the four base blocks are systematically increased?“

The instruction scene which is thoroughly analyzed in this section originates from the third lesson of the teaching sequence held by teacher B on the topic of number walls. The entire episode „How does the goal block change when the four base blocks are systematically raised?“ comprises the last 19 minutes of this lesson. In this episode, the conversation is first about the exercises that the children had been working on with a partner or in small groups. They were supposed to increase the four base blocks of a given number wall by 10 one after another, calculate the new number walls which emerged from this, and talk about the striking features observed.

In the whole class discussion, the student Moritz had set up the following assumptions about the change of the goal block when increasing the base blocks. When a border block (first or fourth block is raised by 10, the goal block is raised by

10 as well. When a middle block (second or third block) is raised by 10, the goal block is raised by 20. In their work on number walls, however, children have observed that the goal block is raised by 30 when a middle block is raised by 10. In order to be able to explain and justify this observation in an easier way, the teacher goes on to a new, easier to calculate number wall, in which there are only tens in the base blocks. In this number wall, systematic changes in the base blocks by 10 are carried out as well. On the board, one can see a finished, calculated number wall at the top and two number walls at the bottom, in which one of the base blocks has been changed (see Fig. 77).

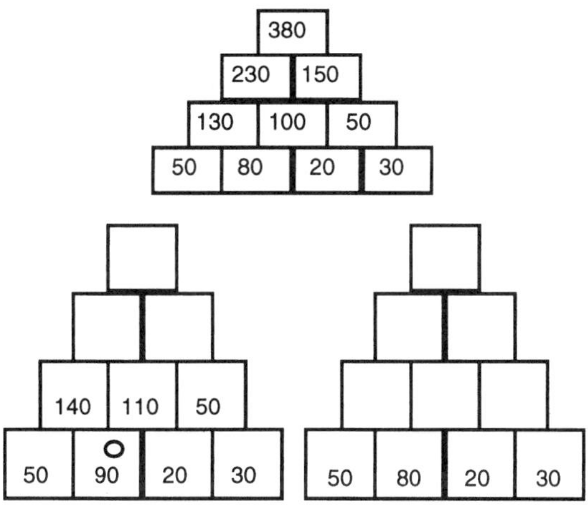

Figure 77. The change of a base block in a number wall.

With the lower left wall, the teacher has marked the 2nd block with a magnetic chip in order to point out that this block has been increased as compared to the top number wall; the student Hannah has calculated the second row of this number wall. At this point, the teaching scene starts with the teacher's question: „What has changed? By how much?".

165 T And perhaps already why as well! Monika!

166 Mk [*goes to the black board*] That here has become a hundred and forty [*points at the fifth block of the lower left number wall*] and that a hundred and ten [*points at the sixth block*], and up there it was a hundred and thirty and one hundred [*points at the upper number wall*], and ehm [*points at the second block of the lower left number wall*], because that is ten more now [*points first at the seventh and then at the fifth block*], ehm, here. [*points at the second block*]

167 S Ah yes!

168 Mk Because that is calculated twice [*points first at the third and then at the fifth block*], that is twenty more above too- [*points at the tenth block*]

169	S	I also know why.
170	T	We'll come up there soon. [*puts a chip into the fifth block*]
171	Mk	... and, ehm [*points at the fifth and the sixth block at the same time*], then the ninety [*points at the second block*] is twice. Here [*points at the third block and then at the first block*] ... calculated plus fifty and then plus twenty. [*points at the third block again*]
172	T	[*puts a chip into the sixth block*] ... Yes. ...

Summing up, this short scene can be described in the following way. First Monika names the numbers which have changed. She points at the 5th and 6th block in the left wall and says: „That here has become 140 and that 110". She points at the upper wall: „... up there it was 130 and 100". Then she starts with an explanation. She „searches" for the 2nd block of the left wall and states: „... because that is 10 more now. ... Because that is calculated twice, that is 20 more above", and she points at the goal block. The teacher puts a chip on the 5th block. Monika points at the 5th and the 6th block in the second row again, points at the 2nd block: „... then the 90 is twice. Here ... calculated plus 50", and she points at the 1st block and then at the 3rd block „plus 20". The teacher puts a chip into the 6th block; she confirms it and requests Monika to look at the next row.

The knowledge construction which is communicated by Monika by means of speaking and pointing at positions in the number walls can be represented with the following signifiers. The diagram with the three number walls on the board (Fig. 77) represents the signified or the reference context. In her contribution, the student Monika refers to two number walls, the one above and the lower left one.

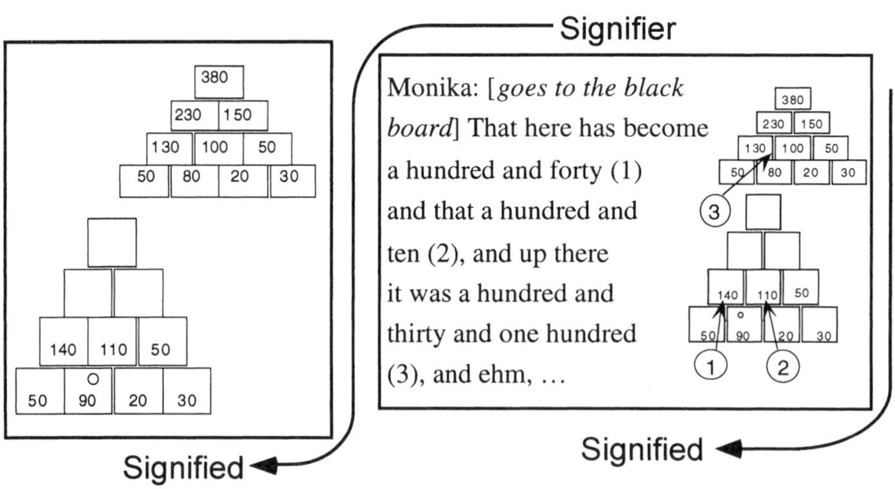

Figure 78a. Monika's signifiers to justify the increase in the goal block (1. part).

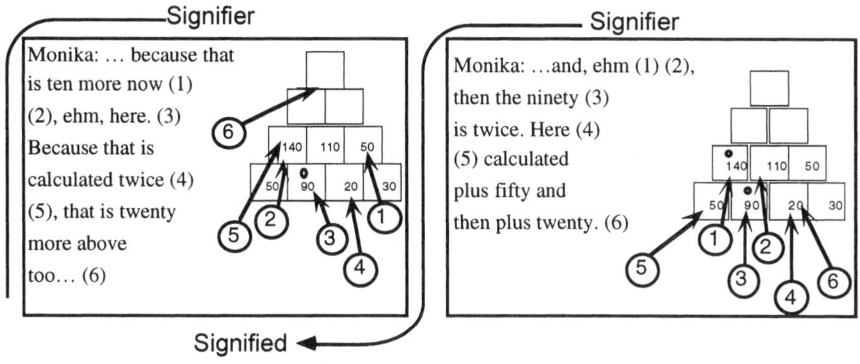

Figure 78b. Monika's signifiers to justify the increase in the goal block (2. part).

Monika constructs a first signifier by means of speaking and pointing at: The numbers of the 5th and 6th block have become 140 and 110 as opposed to the blocks in the original wall. Then she begins to formulate and to show a second, justified signifier: She points at several blocks (7th block, 5th block, 3rd block), emphasizes that there are 10 more, and she adds „Because that is calculated twice (points at the 3rd block at the same time), that is 20 more above". By saying this, she intends the relation: „An increase in 10 is calculated twice, thus the number above is raised by 20". With her signifier, the teacher indirectly restricts the observation to the lower blocks of the wall: „We'll come up there soon." Furthermore, the teacher puts a chip into the fifth block which is supposed to suggest the increase by 10. Thus Monika develops a third signifier to explain the effects of the 2nd block on the second row of the wall; when speaking she points at the individual blocks of the wall again – partly searching for the block she is speaking of: »... the 90 is twice. Here ... calculated plus 50 and then plus 20." The consequence on the 5th and 6th block which results from this remains unsaid, but it is potentially contained in this signifier against the background of the calculation rules for number walls. The teacher confirms her explanation and attaches another chip.

First, Monika wanted to infer directly from the operational connection of the „double calculation of the 2nd block" (which had already been stated in the previous questions) to the increase in 10 of the block above (without paying attention to the individual transitions in the rows of the wall). The teacher restricts the observation to the first two rows. Then Monika gives a justification for the fact that the increases in the 5th and 6th block by 10, which she had mentioned at the beginning, result from the „double calculation" (of the 2nd block) (without speaking about that directly here). By means of the chips which are put on the 5th and the 6th block by the teacher, another signifier („mathematical sign") for this is given.

From an epistemological perspective, Monika constructs a new knowledge relation by means of trying to deduce a justification for the increase of the goal number from the „double calculation" of the 2nd block. She employs the increase of the 2nd block by 10 as well as the fact that this block is calculated twice. On this basis she infers that the uppermost block is raised by 20 (two times 10). In a second consideration, she explains how the second block with the two blocks beside it, is calculated twice, without mentioning that the 5th and 6th block are thus increased by 10. It remains open in which way the increase in the goal block (by how much) can be explained by the „double calculation" of the 2nd block. The new knowledge relation which is intended by Monika can be represented in the epistemological triangle as follows (Fig. 79).

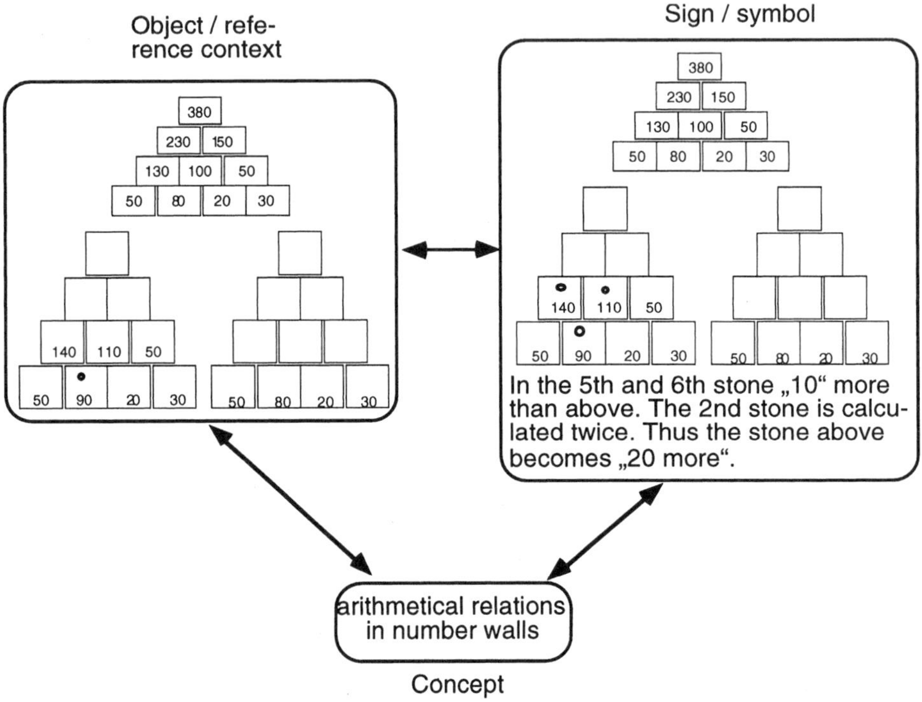

Figure 79. The epistemological triangle: Monika's explanations of the increase in the 5th and 6th block as well as in the goal block.

With this relation the increase of the block „above" (goal block) by 20 ($2 \cdot 10$) is directly inferred from the increase by 10 of the 2nd block (an inner block). Furthermore it is explained how the number on the 2nd block appears twice in addition exercises. This relation between the inner block (2nd block) and the increase of the goal block remains open; there is a lack of explainatory steps. The connection between „the raised number is calculated twice" and „thus the goal number is raised by 20 ($2 \cdot 10$)", is formulated as a justification, but it is not established as a consistent connection with the construction relations for number walls (this connection is not even mathematically correct).

The communicational and epistemological analyses of Monika's knowledge construction make it clear that elements of factual knowledge are contained in the establishment of the magnitude of numbers as compared to the original wall. Monika then shows and formulates a justification which is based on the relations in the present number walls and which infers the change in the goal block directly from the observable increases. Thus a condition for the construction of a mathematical sign in the mediation between reference context and sign / symbol is fulfilled in a first approach, namely that mathematical signs are about relational structures. The second condition for a mathematical sign is not satisfied, namely it is not mentioned that the stated relation for the change in the goal block is founded on the presupposed conceptual construction relations for number walls and that it is compatible with these. Thus Monika does not construct a true mathematical sign as a justification for the question of the amount by which the goal block changes when a middle block is increased by a fixed value.

3.3 Teaching Episodes from Teacher C's Instruction

3.3.1 Analysis of a Scene from the Episode „How can the 20th triangular and rectangular number be determined?"

In this section I will use the communicational and epistemological perspectives to examine a classroom scene from teacher C's class. The mathematics teaching in this class was based on a learning environment about „figurate numbers". In the first two lessons of the instructional sequence, the 3rd grade students had been acquainted with rectangular and triangular numbers. Working in groups, children produced posters with at least the first ten rectangular or triangular numbers using dot patterns and corresponding sums and products.

In the third lesson of this short sequence, children were working on different „investigations" set by the teacher individually or with partners nearly all the time. The following episode shows the beginning of the discussion of these investigations. The discussion was taking place in the so-called theater circle setting and had begun in the last minutes of the lesson. In this episode, the student Jan-Pattrick explains and justifies his way of determining the 20th triangular number. The second, short episode originates from the beginning of the following fourth lesson from teacher C's teaching sequence. Jan-Pattrick again explains his way of proceeding to determine the 20th triangular number.

77	T	So ... now you have all been researching very diligently, as far as I could see. And now you still have the possibility, if you want to, to explain to the others whether you have found out anything particular. Have you discovered anything during your work?
77a	SS	[*whispering*]
77b	T	Thorsten, you can raise your hand, then you can say something. Jan-Pattrick.
78	Ja	I have found out the twentieth triangular number.
79	T	Well. ... Can you <u>reveal</u> it to us?
80	Ja	Two hundred and ten.
81	T	Mhm. ... Perhaps you can write it on the board there, then we have got a result there. ... [*Jan-Pattrick goes to the board and writes down „210"*] Can you also explain how you found it out?

82 Ja First I have always drawn a triangle, thus, the pictures. Then I have always written the sum.

Picture	Addition Task	Triangular Number
	0 + 1	1
	1 + 2	3
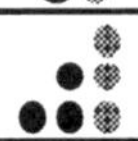	3 + 3	6
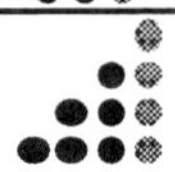	6 + 4	10
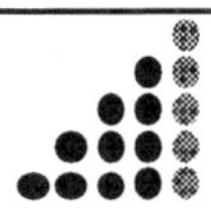	10 + 5	15
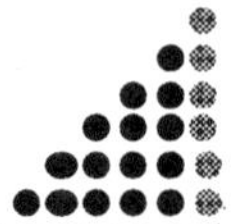	15 + 6	21
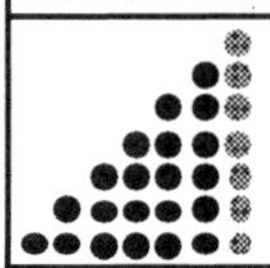	21 + 7	28
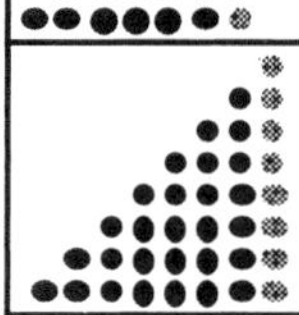	28 + 8	36
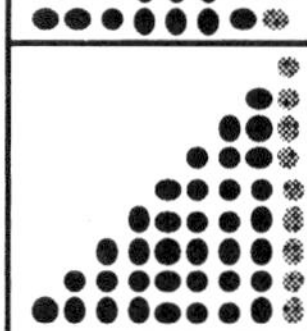	36 + 9	45

Figure 80. Geometrical and arithmetical connections between triangular numbers.

83 T And which sum have you written, can you tell us more precisely?

84 Ja [*goes back to his seat*] Well, the thir… the fourteenth was, I believe, ninety-one plus four first calculated because it was the fourteenth. Is ninety-one plus fourteen … [*incomprehensible*]. Then I have, ehm, first plus four, that was ninety-five, then plus ten, then that was one hundred and five.

85 T Mhm. Well, if I understood you correctly, then you always went on calculating this sequence of sums. Right? [*T points to the left of the board to a poster with triangular numbers from top to*

bottom. To the left of it, there is the poster with rectangular numbers (with products and „long“ sums)]

86 Ja Yes.

87 T And then you have finally found out that one. Mhm. Yes, that is already a good story, you have found out. ... Who else has discovered and researched anything?

This short scene can be summarized and structured as a sequence of smaller interaction phases. Children have engaged in investigations for their posters with triangular and rectangular numbers in pair work. Now the teacher wants some of the results of this work to be presented and discussed. He calls Jan-Pattrick.

Phase 1 (78 – 81) Jan-Pattrick names 210 as the quantity of the 20th triangular number

Jan-Pattrick says that he has found out the 20th triangular number and names it in response to the teacher‘s question: „210“. He also writes this number on the board.

Phase 2 (81 – 82) Jan-Pattrick has continued several drawings and the addition exercises

Jan-Pattrick explains how he got this result. First he drew more pictures, then he continued with the sums.
[These were sums for triangular numbers whose 1st term represents the previous triangular number and the 2nd term gives the number of chips which are new. These chips are colored red in the picture and represented as the vertical side of the triangle. The quantity of these new chips gives the number of the respective triangle.]

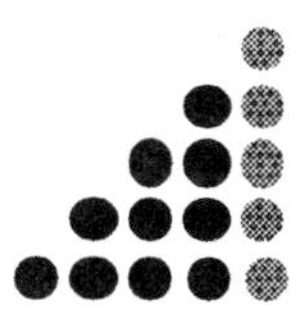

Phase 3 (83 – 84) Jan-Pattrick explains his calculation at the example of the 14th triangular number with the addition exercise 91 + 14 = 105

Jan-Pattrick explains the sum he used more precisely. He refers to the „fourteenth“ sum. He remembers that the first term of the sum was 91, and he wants to add „by positions“: 91 plus 4, because „it was the fourteenth“. First, he names the complete sum again: „91 + 14“, and then he repeats the steps of the addition: 91 + 4 = 95, then + 10, which comes to 105.

The teacher summarizes (85 – 87) Jan-Pattrick‘s way of proceeding as the continuation of the sequence of sums corresponding to the triangular numbers. He points from the top to the bottom of the triangle poster with the pictures, the (short) sums and the triangular numbers.

At the beginning of the lesson on the following day, the teacher reminds the students of the fact that Jan-Pattrick has determined the 20th triangular number. He is supposed to explain his way of proceeding once more. To facilitate the presentation, the teacher transferred Jan-Pattrick's worksheet onto a poster in an enlarged form and attached it to the board (Fig. 81).

The opposite poster is attached to the board. The first two numbers of every row and the plus sign are written in blue, the numbers in front of the equal sign and the equal sign itself is red and the last number in each row is green.

13. 91
14. 91 + 14 = 105
15. 105 + 15 = 120
16. 120 + 16 = 136
17. 136 + 17 = 153
18. 153 + 18 = 171
19. 171 + 19 = 190
20. 190 + 20 = 210

Figure 81. Sums for triangular numbers.

1	T	...Can you explain that to us again, Jan-Pattrick?
2	Ja	Well, with the thirteenth, eh, well, with the fourteenth sum, with the thirteenth, ninety-one was the result, with the fourteenth it was then ninety-one, I have written there plus four, that is, then I have written, equal to ninety-five, then plus ten is equal to a hundred and five. And so on, always.
3	T	Mhm. [*6 sec pause*] What have you done before, with the thirteenth, how did you get there?
4	Ja	Before I have always, ehm, drawn the triangular numbers as you have it there on your poster [*incomprehensible*].
5	T	Mhmm. [*3 sec pause*] Ah yes. [*6 sec pause*] Good, and so, how have you done it then, found out the 20th? ...
6	Ja	Mh, yes, I have then calculated a hundred and ninety plus the twentieth and then I have got two hundred and ten as the result.

This short scene can also be summarized in a sequence of three sub-phases.

Phase 1 (2) Jan-Pattrick explains the calculation of the 14th triangular number with the help of the 13th triangular number

Jan-Pattrick uses the result of the 13th triangular number „91" and then calculates in steps „plus 4 is equal to 95, then plus 10 is equal to 105". He adds: „And so on, always."

Phase 2 (3 – 5) The 13th triangular number is determined by means of continuing the pictures

Jan-Pattrick says that, to determine the 13th triangular number, he drew the triangle pictures first.

Phase 3 (5 – 6) The 20th triangular number results from 190 plus 20

Jan-Pattrick describes how he obtained the 20th triangular in the following way: 190 plus the twentieth gives 210. (He does not say here: The 19th triangular number, 190, plus 20 gives 210.)

The knowledge construction to determine the 20th triangular number which Jan-Pattrick explains essentially by means of speaking can be represented with the following signifiers. The poster on the wall (Fig. 80) with geometrical figures and arithmetical numbers and exercises as well as the sequence of sums (Fig. 81) represent the signifier or the reference context used by Jan-Pattrick.

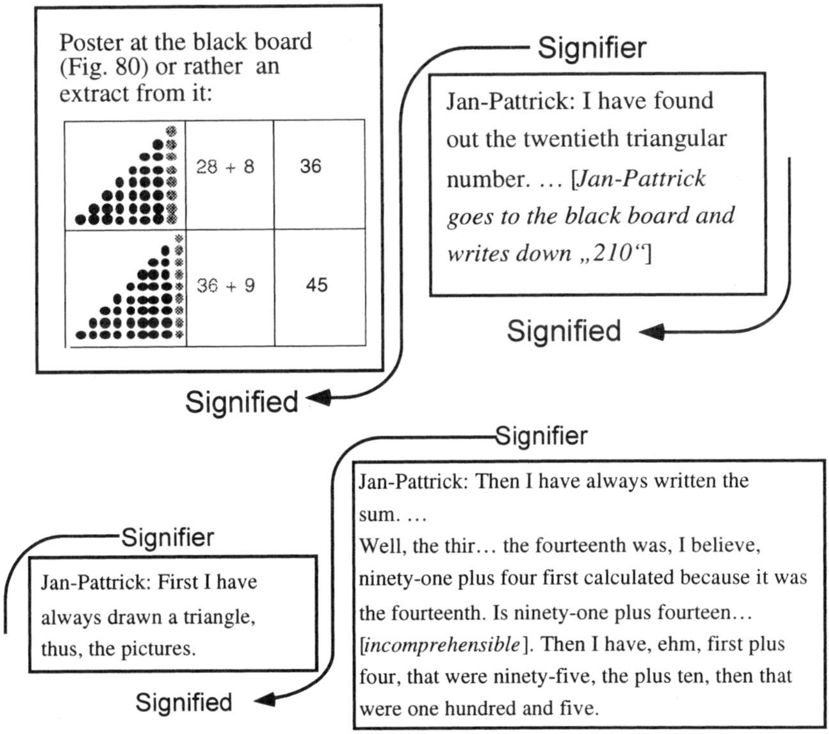

Figure 82. Jan-Pattrick describes his way of proceeding to determine the 20th triangular number in a chain of signifiers.

In his first signifier, Jan-Pattrick communicates that he has found the 20th triangular number and writes it on the board: 210. With the second signifier, Jan-Pattrick intends a kind of continuation of the geometrical patterns that appear on the poster. He does not say concretely how he has determined the respective triangular numbers this way; one can suppose that he has somehow counted these numbers from the dot pattern.

In the third signifier, Jan-Pattrick only refers to the arithmetical aspects of the question (perhaps it was no longer possible or too complex to draw the „pictures"). He says that he has always written down the sums, similarly to those on the poster. Then he once speaks of the „fourteenth" (sum?) and also of 14 as a term of a sum. „It is ninety-one plus fourteen". This can implicitly mean the connection between the 14th sum and its term 14. After that, Jan-Pattrick describes how he has calculated the sum 91 + 4 + 10 = 105 step by step. The teacher interprets Jan-Pattrick's explanations globally as an attempt to continue the sequence of sums.

Then Jan-Pattrick repeats his way of proceeding to determine the 20th triangular number in a sequence, in a way which is a little different.

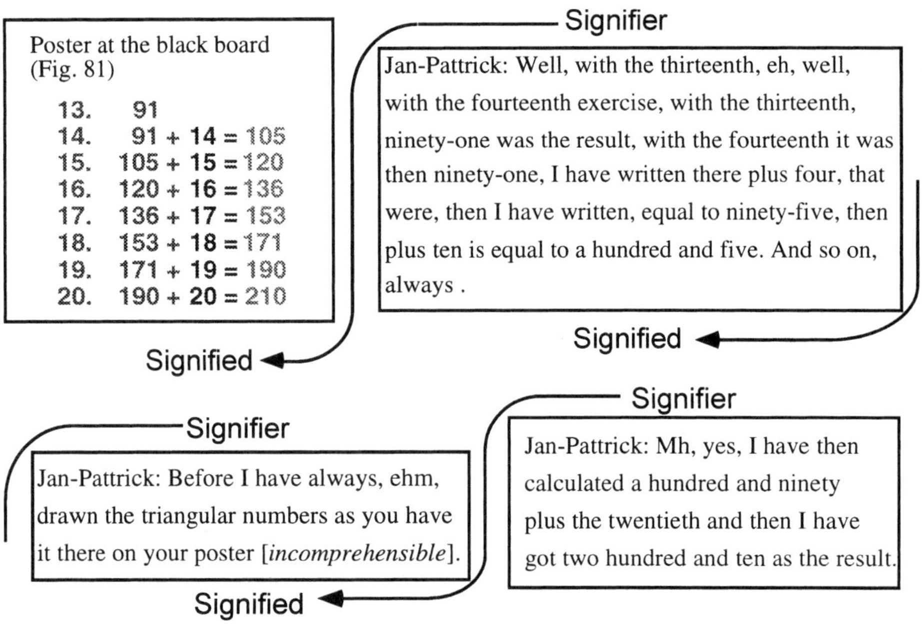

Figure 83. Jan-Pattrick describes his way of proceeding to determine the 20th triangular number in a further chain of signifiers.

In his explanation which he repeats at the beginning of the next lesson, Jan-Pattrick formulates signifiers in response to the teacher's questions. In his first signifier, he refers to the fact that the result 91 was calculated in the 13th sum. Then he names the calculation of the 14th sum on the poster: 91 + 4 + 10 = 105. Furthermore, he suggests a possible continuation, which shows how the further sums are calculated, with the words „And so on, always." Jan-Pattrick does not merely read the sum, he links it to the result of the previous sum.

With his question: „ … with the thirteenth, how did you get there?", the teacher confronts Jan-Pattrick's explanation of the 14th exercise with the result of the 13th exercise whose calculation or determination cannot be interpreted on the poster. Jan-Pattrick answers with his second signifier which refers to another poster and thus assigns another way of producing the 13th triangular number: The drawing of the corresponding „pictures“ and probably also of the corresponding (short) sums.

The teacher‘s question after the 20th triangular number makes possible a reference to the immediate problem with the short sums again. Jan-Pattrick only states in his third signifier that he has calculated 190 plus the twentieth. In fact, he has calculated „190 plus 20“, but calls the second term of the sum „twentieth“, which could hint at the particular item that one calculates „plus 20“ for the 20th triangular number.

From an epistemological perspective, Jan-Pattrick constructed a mathematical knowledge relation by means of explaining a sample calculation and then making it clear that this calculation was meant to refer to a more general procedure. In the first scene, Jan-Pattrick gave first indicators of a possible, developing explanation for the teacher‘s question, how one can determine the 20th triangular number in continuation of the previous work on triangular numbers and short sums. He calculates this 20th number correctly as „210“. First, Jan-Pattrick said that he drew further pictures (dot patterns for triangular numbers) in going constructively from one triangular number to the next. After that, he said, he wrote sums (short sums). Then he probably named the first sum which he calculated without a picture. He went back to the 14th (sum). From his interpretation, a very open, possible (important) connection between the 14th sum and the addition of 14 in this sum becomes visible, which is yet not spoken of explicitly. This is where a possible new construction of mathematical knowledge could lie. In his calculations, Jan-Pattrick must also have used this construction systematically without necessarily having to make it explicit. He could also have transferred it schematically since he determined the 20th triangular number correctly after all. The teacher summarized Jan-Pattrick‘s procedure globally as a continuation of the way of proceeding which was started for the determination of short sums on the poster.

In his second explanation, Jan-Pattrick repeated and confirmed his procedure to determine the 20th triangular number. On the one hand, he said that he had continued the pictures first, and this is probably how he obtained the „result“ of the 13th triangular number. He explained the calculation of the 14th triangular number from the „result“ of the 13th triangular number once more. Here he calculated: $91 + 4 = 95$ and $95 + 10 = 105$. For the twentieth exercise he then explained: 190 plus the twentieth, resulting in 210. Even though Jan-Pattrick did not refer explicitly to the connection between the previous and the following sums in his statements and explanations (by saying, for example, that the previous result is the first term of a sum in the following sum, and that the number of the sum determines the second term of the sum), one can interpret his interpretations so that he uses these relations when continuing the calculations probably in an automatic way. He named two sample calculations concretely, in which, however, open, generalizable interpretations in the interaction are possible.

In the epistemological triangle, the knowledge construction which was observable in the two scenes can be represented in the following way. Jan-Pattrick describes a calculation procedure using concrete numbers and operations. This procedure is intended to be interpreted as a general procedure, and the particular item is that one must calculate the „result“ of the previous triangular number plus the 14th or the twentieth in order to determine the 14th (20th) triangular number.

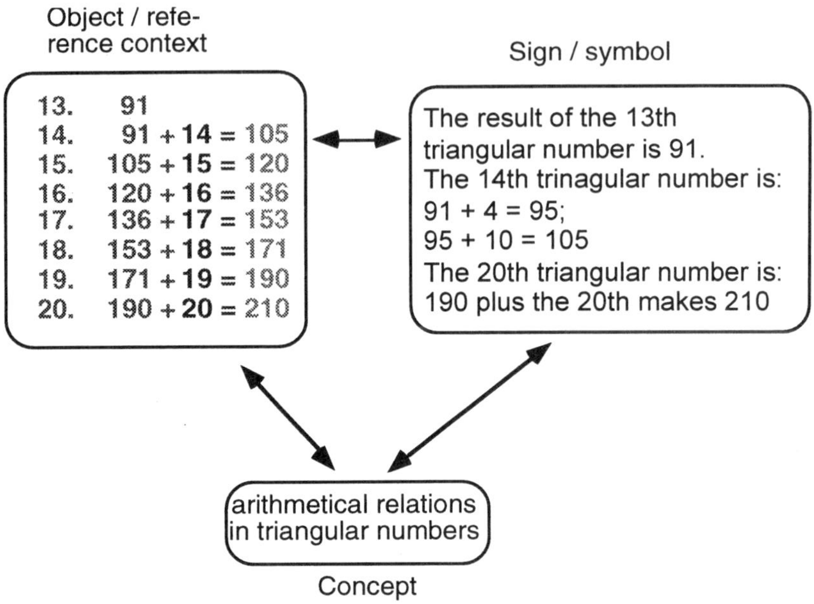

Figure 84. The epistemological triangle: Jan-Pattrick's exemplary procedure to determine the 20th triangular number.

Jan-Pattrick describes a knowledge relation between the successive calculations of the triangular numbers guided by the developed sums and the observable arithmetical structures in sample sums (91 + 14 = 105 and 190 + 20 = 210). This construction is intended to be general. Furthermore it contains examples of relations between the number of the sum and the second term of the sum. Thus, relational arithmetical aspects are essentially contained in the procedure. On the other hand, it is remarkable that Jan-Pattrick carries out the calculation of the 14th triangular number with a concrete supporting task. The knowledge relation developed by Jan-Pattrick is based only on the arithmetical addition exercises; the geometrical structures and possible general relations in the dot patterns are not employed for the description of the knowledge relation.

The knowledge relation constructed by Jan-Pattrick is founded on the description of a procedure for the continued determination of triangular numbers. In his verbal description of the procedure, generalizable intentions can be observed. Then arithmetical relations of the sums – partly on the apparent surface of the arithmetical signs – are used in some aspects. Jan-Pattrick does not make use of any structural aspects of the geometrical triangular patterns to construct the mathematical knowledge, but he confines himself to the arithmetical structure. The mediation of his verbal sign to describe a procedure in the reference context of the addition exercises used by him thus only represents one feature for the construction of a true mathematical sign. Relations are constructed between structures and not only mainly between visible features. However, in this problem connection no conceptual, elementary knowledge relation for the continued construction of the geometrical triangular patterns is used as the starting point to construct this mathematical sign. Thus the second important epistemological condition to construct actual mathematical signs is not satisfied.

3.3.2 Analysis of a Scene from the Episode „What are the connections between rectangular numbers and triangular numbers?"

The classroom scene which will be analyzed in this section is taken from the fifth and last lesson carried out by teacher C based on the learning environment of „figurate numbers". In the second half of the previous fourth lesson, the question which rectangular and triangular numbers can be discovered in a 4x5 dot field was worked on individually or in groups. In the subsequent whole class discussion, Jessica pointed out that the fourth triangular number is „hidden" twice in a 4x5 dot pattern, and another student stated that the triangular number represents exactly one half of the dot pattern. Already at this point, the teacher remarked that Jessica's discovery is connected to the „big secret" which links rectangular and triangular numbers and which would be examined more closely on the following day.

The activity with this „secret" then fills out the complete fifth lesson of teacher C's instruction sequence, and the selected episode „What are the connections between rectangular numbers and triangular numbers?" takes up about the first 17 minutes of it. Sitting in the theater circle in front of the board, children investigate the question of whether a triangular number is always half of a rectangular number or whether a rectangular number is always double a triangular number during the whole episode. This connection is checked and confirmed on the basis of the example of the first seven rectangular numbers, and then it is considered as to how one could find out whether it is always valid. The episode ends with the teacher asking children to go back to working individually or with partners further on the question of the generality of the discovered relation.

In the following scene, all children are sitting in the theater circle facing the board. Displayed on the board are dot patterns for the first five rectangular numbers (divided into two triangular configurations by means of different colors) together with the values of the respective triangular or rectangular numbers (see Fig. 85). The discussion now focuses on determining the values and the configuration for the 6th position.

88	T	Yes. So what can we do to find out if this is always true?
89	S	Nothing.
90	T	Christopher.
91	Ch	I notice something.
92	T	Yes, tell us.
93	Ch	Up there it goes four. Then it goes six. Then it goes eight. And then it goes ten. [*At this moment, T points at the number 20 and then at the number 30 on the left hand side of the table*]. Then it goes twelve [*T now points at the empty field below the number 20*]. Therefore there should be thirty-two on the other seventeen [*2 sec pause*] um, forty-two should be on that and on the other one twenty-seven

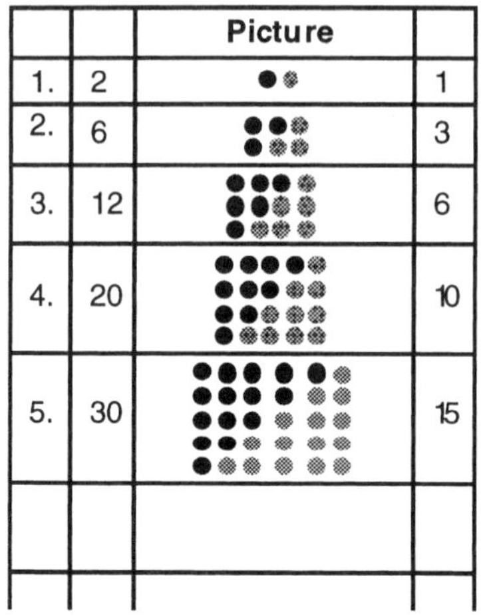

Figure 85. The connection between triangular and rectangular numbers.

94	T	I see. You mean..., that's quite an interesting idea, Christopher. You mean, here there should be forty-two? [*points at the empty field below the number 30*]
95	Ch	Yes. [*T writes the number 42 in the table*]
96	T	Yes. And there? [*points at the empty field below the number 15 on the right side of the table*]
97	Ch	Twenty-seven.
98	T	Why do you think there should be twenty-seven? ... Can you give a reason for that?
99	Ch	No.
100	T	No? ... Nico.
101	N	Twenty-one.
102	T	Why do you think [it's] twenty-one?
103	N	Because twenty and twenty are forty [*points at the ten's decimal place of the number 42*] and one and one are two [*points at the unit's place of the number 42*].
104	T	Mhm [*writing the number 21 in the table*]. Oh yes, then we already know the next thing. But we ought to check whether it is indeed correct from the picture, whether it is really always like this.

		Picture	
1.	2		1
2	6		3
3	12	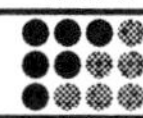	6
4	20	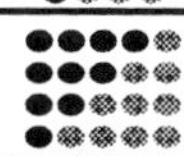	10
5	30	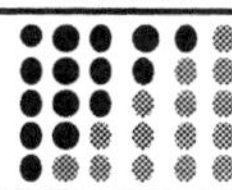	15
	42		21

Figure 86. New triangular and rectangular numbers.

This classroom scene will first be structured and summarized. First the teacher again asks his question as to whether it is always true, and after that, he asks : „How can we find out whether this is always true?"

Phase 1 (90 – 97) Christopher continues the numbers in the column of the rectangular numbers and derives a new triangular number

Christopher noticed something. He names the sequence of numbers one after another: „... there it goes 4, then it goes 6, then it goes 8, then it goes 10, then it goes 12". With this, he seems to refer to the second column, and the teacher points at this column, at the numbers 20 and 30. Christopher named the respective difference or increase between the numbers in his sequence. Then he infers: „Therefore there should be thirty-two and on the other seventeen". He has (mistakenly?) constructed a number bigger by 2 in the left number column, and he does the same in the right number column: from 15 to 17. Christopher corrects his statement: „Forty-two should be on that and on the other one twenty-seven." Here he has raised the two numbers by 12. The teacher confirms the first number with the question whether „here there should be forty-two?" and he writes this number down after Christopher has agreed. Christopher repeats once again that, in the other position, there should be „27".

Phase 2 (98 – 100) Christopher cannot justify his procedure

The teacher asks Christopher to justify his claims. But Christopher cannot justify why „27" is supposed to be here.

Phase 3 (100 – 104) Nico corrects Christopher's triangular number and gives a justification for his claim

Nico says „21" and means that this number is correct. He justifies this with the following „calculation": „Because twenty and twenty are forty [*points at the ten's decimal place of the number 42*] and one and one are two [*points at the unit's place of the number 42*]." The teacher agrees with him and writes down the new numbers (see Fig. 86). The teacher formulates a new „research mission": „But we ought to check whether it is indeed correct from the picture, whether it is really always like this."

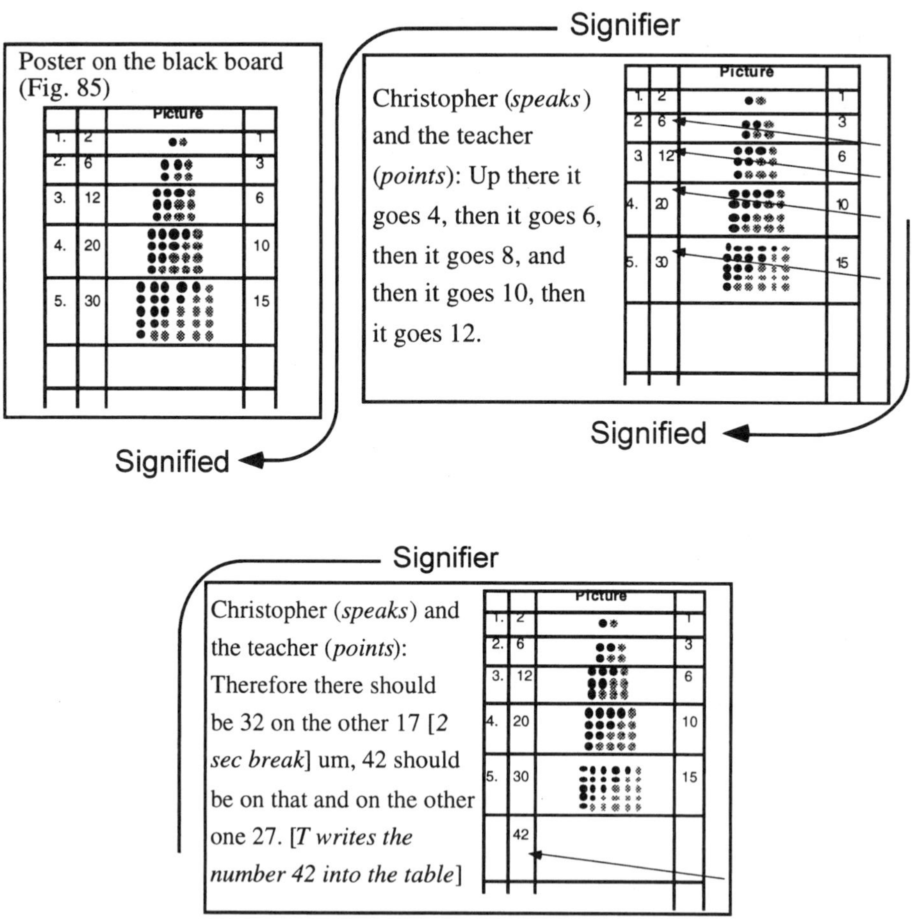

Figure 87. Christopher explains the continuation to the new rectangular and triangular number in a chain of signifiers.

First, Christopher's contributions (together with the teacher's pointing gestures) can be represented in a sequence of signifiers (Fig. 87). With the sequence of the numbers: „Up there it goes 4, then it goes 6, then it goes 8, then it goes 10", Christopher names the differences between the rectangular numbers which have been written down. The teacher points at the respective number column. With his description, Christopher seems to have in view the additive continuation or building of the number sequence. He continues this signifier: „Then it goes 12"; this increase by 2 is supposed to lead to the new rectangular number in the 6th row.

Christopher now uses his arithmetical progression as a justification for the new numbers. He infers first the two numbers 32 and 17, numbers which differ by 2 from their antecedent numbers – perhaps Christopher transfers the increase of the differences directly to the new situation. He corrects himself immediately. Now, however, he seems to add 12 – the new difference – respectively in both cases, and he names the numbers 42 (rectangular number) and 27 (triangular number).

In his argumentation, Christopher referred only to the – rather complicated – arithmetical continuation pattern of the rectangular numbers, he did not take the geometrical situation into consideration. He applied this principle to determine the rectangular number 42 – which is confirmed by the teacher –, and he transfers the principle also to the sequence of the triangular numbers and thus determines the sought for triangular number as 27 (as 15 + 12). After the teacher's question, Christopher confirms that he believes that 27 belongs in the empty field. However, he cannot justify his claim.

Nico continues the knowledge construction. He formulates a signifier as justification for the choice of the number 21 (Fig. 88). His justification „Because 20 + 20 are 40 and 1 + 1 are 2" results in a dissection of 42 into 21 + 21 or of 40 into 20 + 20 and of 2 into 1 + 1. If this is put in connection with the relation „Always half" (the triangular numbers are always half of the rectangular numbers), which has been thoroughly discussed before, Nico intends a justification by using this relation. The teacher confirms this correct number by writing down 21.

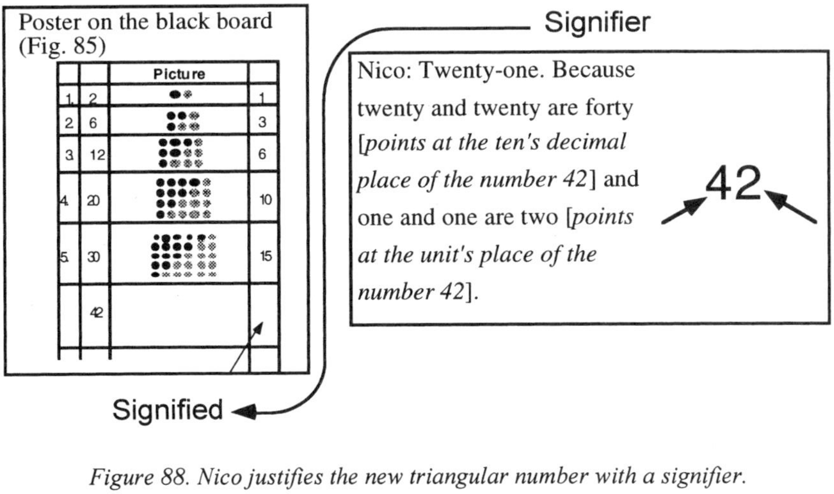

Figure 88. Nico justifies the new triangular number with a signifier.

From an epistemological perspective, the knowledge constructions of the two students can be characterized in the following way. One can represent the mediation between sign / symbol and reference context carried out in this situation by the following case of the epistemological triangle (Fig. 89).

Concerning Christopher's knowledge construction, the analysis shows that he developed a continuation principle for the 6th rectangular number from the given arithmetical pattern. His counting by twos, 4, 6, 8, 10, 12, is meant to suggest that the difference between the rectangular numbers is always an increase by „2", and that therefore „12" must now be added to the value of the 5th number. This addition of „12" is transferred to the 5th triangular number, and „27" is determined as the 6th triangular number. Christopher constructs a general arithmetical relation between the rectangular numbers in a verbal way and transfers it directly to the triangular numbers. This connection is inferred only from the arithmetical structure. No justification is given, for instance using the geometric pattern of the rectangular numbers.

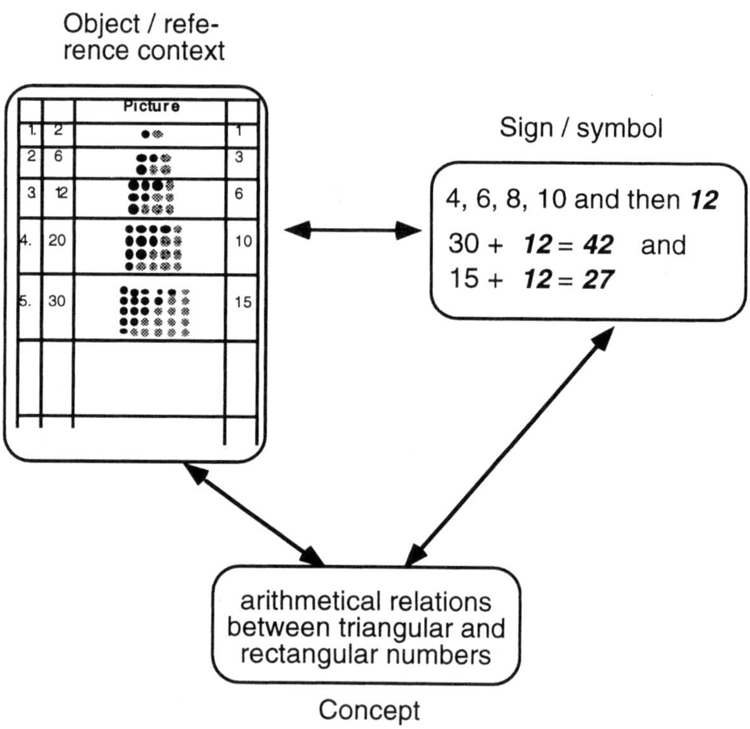

Figure 89. The epistemological triangle: Christopher considers arithmetical distances of number sequences.

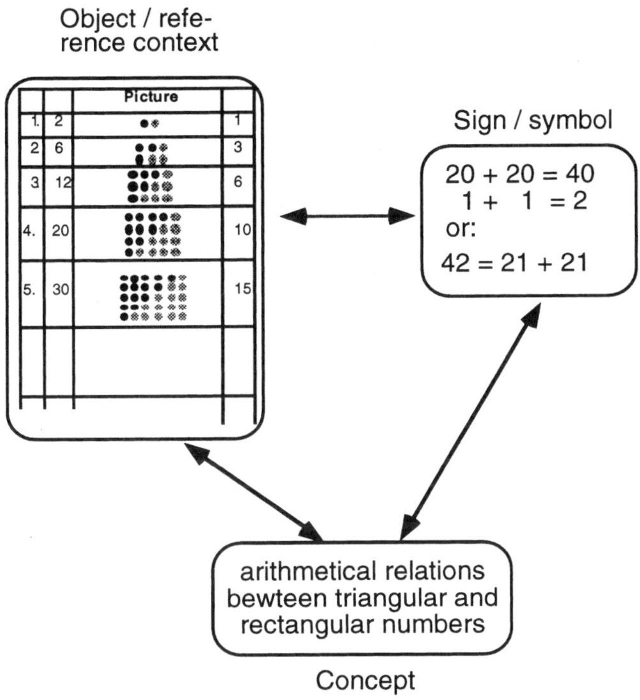

Figure 90. The epistemological triangle: Nico dissects 42 into 20 + 20 and 1 + 1.

In Nico's knowledge construction, the rectangular number „42" is halved in an particular way, first the ten and then the one. The intention connected with that – the triangular number is half of the corresponding rectangular number – is not articulated directly. The brief argument is restricted to the procedure of the arithmetic bisection or doubling only. From an epistemological point of view, Nico constructs a brief, verbally formulated sign „20 + 20 = 40 and 1 + 1 = 2" with reference to the number 42 which was noted on the poster. This mediation between sign / symbol and reference context is represented as above in the epistemological triangle (Fig. 90).

In their contributions, both students constructed knowledge relations that were new; they could not be directly inferred from knowledge that was already there. They constructed new arithmetical, structural connections, which were not used or recognized in this form before. It is striking that these knowledge relations were restricted to the arithmetical number symbols and structures with no reference to the geometrical configurations.

The question why the structure which was observed locally in the numbers was really generally valid and „always" continuable could be answered by reference to the geometrical, general connection between triangular and rectangular configurations. The new knowledge relations that were constructed were basically detached from the geometrical problem; it appears that these relations referred only to purely striking arithmetical features and structures.

Based on this analysis, it can be stated that Christopher and Nico construct new relations in the arithmetical context, but their knowledge relations do not refer to the principles of geometrical construction of triangular and rectangular numbers or to the connections between the two configurations. Thus, again, only one of the conditions for mathematical signs is identifiable in their knowledge construction. They construct mathematical signs that are not connected to the presupposed problem knowledge, but signs that use the visible arithmetical structure of the numbers on the poster.

3.4 Teaching Episodes from Teacher D's Instruction

3.4.1 Analysis of a Scene from the Episode „Continuing triangular numbers"

The episode „Continuing triangular numbers" which is observed here comes at the beginning of the teaching sequence by teacher D. She opens this first of four lessons in the learning environment „Figurate numbers" in her 4th grade class by forming two configurations (intended to represent the first two triangular numbers) on the board with magnetic chips and asking the children to form the next pattern in the sequence that she has begun. In response, children present different ideas which could all count as a correct continuation of the given configurations, but which are all rejected by the teacher as not convenient. Only when the configuration corresponding to the third triangular number is finally formed, does she accept the solution. A student now tries to justify why this is the right solution. The course of this classroom interaction until the beginning of the student's justification can be summarized as follows.

The teacher puts two chip configurations on the board.

Figure 91. The first two configurations.

How does it go on? She emphasizes that special numbers are to be considered which have something to do with the chips. Dennis continues the chip pattern in the following way.

Figure 92. Dennis continues a pattern.

This suggestion indeed contains a correct continuation in which the hook is extended by one chip on each arm. The teacher questions this suggestion: „Is this already correct? ... One could think so, but it is not yet completely correct" (5). She refers to the figure and surrounds the first two configurations by a right-angled triangle border.

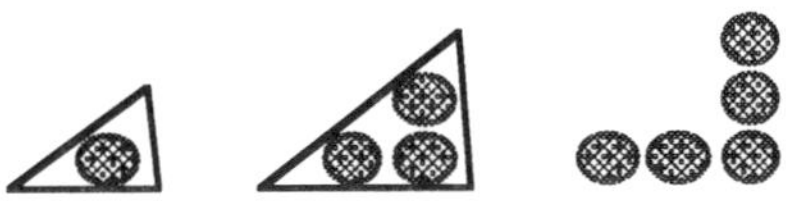

Figure 93. The teacher surrounds two patterns with a triangular border.

The teacher points at the pattern formed by Dennis and says that one could not draw a triangle as a border: „One could obviously not do that here." (7). She goes around the third figure with her hand, suggesting the form of an angle or a hook. The teacher probably proceeds on the assumption that only one single chip in the correct spot is missing for the right solution, and she tries to make the students aware of this in a relatively direct way. With her question: „Who can still form this like that, or use I don't know what?" (7), the teacher expects the missing chip to be put in now. But Lisa makes a new suggestion and forms the following pattern.

Figure 94. Lisa's pattern.

She seems to see the original figure as a whole or one single configuration and looks for a continuation of it. Her suggestion also represents a plausible continuation in which the left part of the figure always receives a further chip horizontally and a chip is added vertically to the right part of the figure. The teacher rejects Lisa's suggestion with the reference to the form of the triangle. At the same time she points at the base and at the hypothenuse of the triangle. Now Kai takes all the chips off that Lisa has put at the board and forms his own pattern.

Figure 95. Kai's patterns.

Kai puts down the last missing chip with his classmates' „support". He has formed an „isosceles" triangle, which is then also emphasized by the teacher. „Yes, now I must have a look here. We had such a form here, but now we have something like that. [*draws a triangle above Kim's configuration*]" (14).

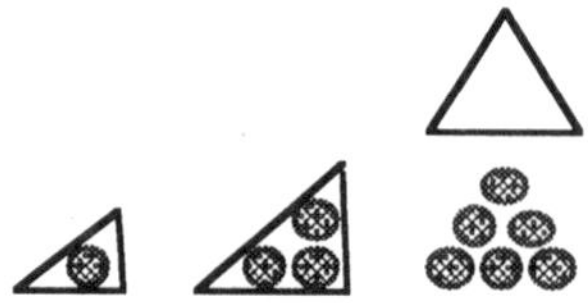

Figure 96. The teacher draws a border.

The teacher questions whether this is the same (16); she thus rejects Kai's suggestion. Kai takes all his chips off the board except for the lower row. The teacher makes him aware of the form of the border once again and says that the new figure has to look like the one that is already there, only with more chips (18). Kai first forms the same figure as in the second pattern.

Figure 97. Kai continues the pattern once more.

Then he adds two chips in the following way.

Figure 98. Kai continues the pattern once more.

The teacher confirms: „Oh yes, he is very close!" (22). Kai's classmates also give hints, that only one more chip is missing. But Kai cannot finish the suggestion. Tugba puts the missing chip in the bottom left spot:

Figure 99. Tugba finishes the new pattern.

Now, let us look in detail at the short scene focused on the justification of the homogeneous character of the pattern.

23	T	…Mhm. Who can explain why is this correct now?
24	T	This is correct. … Rabea.
25	R	Because this is the same pattern again now.
26	T	Mhm. Can you come and draw the pattern around there? … [*Tugba sits down again and Rabea comes to the black board*] But what is different in this one? First make the pattern! Go once around it! [*Rabea draws a triangle around the last configuration*

with chalk, 10 sec pause] Aha. So what is different from this here? [*points alternately at the second and the third triangle*]

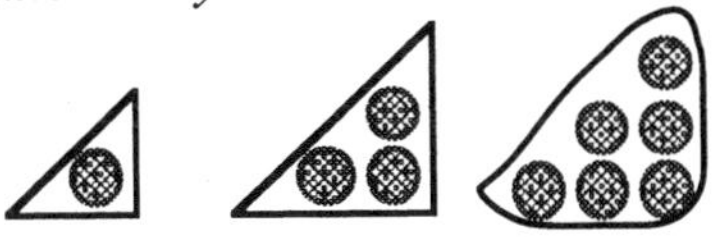

Figure 100. Rabea borders the new pattern.

27	R	There are more chips again. [*points at the third triangle*] But ...
28	T	Yes, where are there more chips?
29	R	At the bottom. And here. [*points at the whole third triangle*]
30	T	Show it exactly! Here is [*points at the lowest row of chips of the third triangle*]
31	R	Here is one more again [*points at the lowest row of chips of the third triangle*] and here is one more. [*points at the upper chip in the second column of the third triangle*]
32	T	There [*points at the upper chip in the second column of the third triangle*]. And where else?
33	R	And here. [*points at the upper chip in the third column of the third triangle*]
34	T	There is also one more. Okay. ...

In answer to the (confirming) question by the teacher: „Who can explain why this is correct now?“ ... „This is correct.“ (23, 24), Rabea answers: „Because that is the same pattern again now.“ (25). Upon the teacher‘s request, Rabea draws the pattern in the form of a triangle around the new geometric configuration (Fig. 100). Furthermore, Rabea is asked to explain the difference between the new, third configuration and the second one. First, she says that there are more chips in the new figure. In response to the teacher‘s question, as to where there are more chips, Rabea answers: „At the bottom. And here.“ and at the same time, she points at the lower and the right row of chips. She is asked to show that more precisely. With the teacher‘s help, the new chips in each row (or column) are pointed at one after another (30 – 33). The teacher expressively confirms the correctness (34).

In a sequence of short verbal statements, by means of bordering the new pattern with chalk and by means of pointing, Rabea justifies why the third configuration is correct. The signifiers stated by Rabea in the interaction with the teacher can be represented as follows (Fig. 101).

In principle, this justification consists of the statement: „Because this is the same pattern again now.“, which is additionally illustrated as the „same pattern“ by Rabea (Fig. 100). How can this statement obtain the status of a justification? On which justifying elements is this argument founded? For this statement, this function of justification is only deducible from the previous form of the interaction. As we have seen, other continuations are indeed meaningful, possible and correct with respect to the problem situation given by the teacher. However, these are rejected by the teacher one after another, while each time she makes the conditions of „homogeneous character" more explicit. Thus alternative possibilities of continuation offered by the children are gradually excluded until only the correct third pattern intended by the teacher is finally produced. Thus, in principle, Rabea has no choice but to directly state the „homogeneous character" of the pattern.

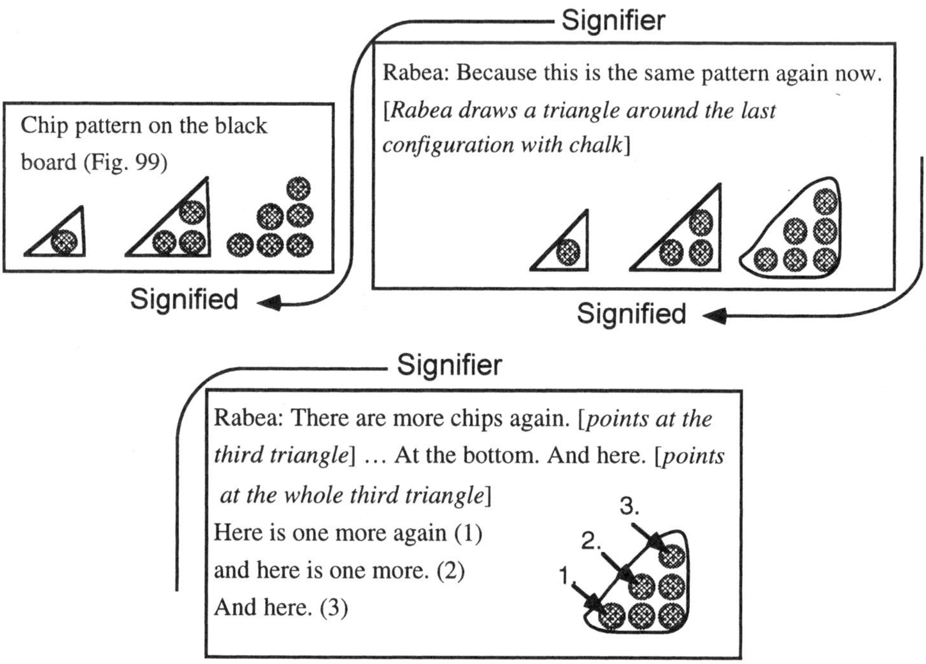

Figure 101. Rabea justifies the continuation of the new triangular pattern in a chain of signifiers.

The more and more explicit assumption underlying the mathematical interaction in this scene is that there is only one correct pattern and thus only one correct third configuration. The teacher emphasizes this several times, and the students accept this interpretation. „Is this already correct? ... One could think so, but this is not yet completely correct" (5); „Mm, is not yet correct like this." (10); „... he is very close!" (22); „... why is this correct now? „ ... „That is correct." (23, 24). From the point of view of communication analysis, the interaction is used to make explicit the relation between the signifier (the two chip configurations) and the „corresponding" fixed signified (the unequivocal figure of a right-angled triangle), which the teacher pre-supposes as correct beforehand.

From an epistemological point of view, it becomes obvious that in this interaction, the relation between „sign /symbol" and „object / reference context" is supposed to have been seen as an unequivocal and fixed connection between the triangular configurations (as opposed to other configurations) and the new chip configurations yet to be produced and this relation is increasingly understood as such. The rejection of the angle pattern (by Dennis, Fig. 92) or the refusal of the continuation pattern „one chip more to the left part of the figure at the bottom left; one chip more to the right part of the figure at the top right“ (by Lisa, Fig. 94) cannot logically be justified or inferred. What intervenes here is a social moment, a convention introduced by the teacher which is supposed to work as the single continuation pattern. The epistemological triangle represents this interpretation as follows (Fig. 102).

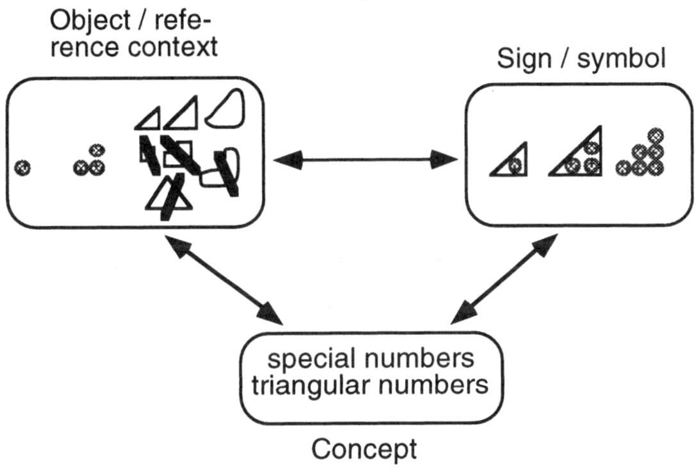

Figure 102. The epistemological triangle: Rabea justifies the triangular pattern.

The new sign and the corresponding configuration are definitely connected beforehand. The signs / symbols become fixed names for existing objects (here fixed configurations). One can observe in the present episode a specific kind of justification and argumentation. To be able to justify the new, yet to be constructed chip configuration (from the two given configurations) as an accepted successor of a former pattern (in the teacher‘s intention as *the* correct one), one has to proceed as follows.

By pointing to differences and similarities between students‘ suggestions and the single pattern the teacher has in mind, in this way she makes this pattern more and more explicit till the students‘ propose a chip configuration which corresponds to it. This continuation pattern is mathematically admissible. However, it is socio-conventionally marked in a particular way in the interaction as the only right pattern. No specific justifications are given for this. It is the teacher‘s authority that comes into play. In the „preparation“ of the argument, one single object (right-angled triangular pattern) is worked out and conventionally marked, which can then serve as the basis of the argument in the direct comparison of the new sign (the third chip configuration) with this interactively reconstructed object (right-angled triangular pattern). If an identity can be stated, then the justification that the new chip field is the correct continuation is acceptable.

The mediation between a new signifier – the third chip configuration – and the reference context of the first two chip configurations with their geometrical borders, which is produced in the common interaction between the teacher and several students in this scene, does not rely on the use of structures and relations. Furthermore, no knowledge relations are applied for the construction of the sequence of triangular patterns. The third configuration is empirically interpreted with reference to the first two configurations and by reference to visible features, namely the correct triangular shape and the three new chips. Thus no full-fledged mathematical sign is constructed in this interaction; the discussion is about conventionalized features and procedures to produce the third configuration.

3.4.2 Analysis of a Scene from the Episode „Connections between triangular and rectangular numbers"

In this section I analyze a scene from teacher D's instruction in the second lesson of a short teaching unit on the topic of „Figurate numbers". At the beginning of the lesson, multiplication tasks corresponding to the pattern of rectangular numbers were discussed. Next, the classroom situation is as follows. Chip configurations for the first four rectangular and the first four triangular numbers are displayed on the board (cf. Fig. 103). Number signs for the corresponding quantities of chips are seen above them.

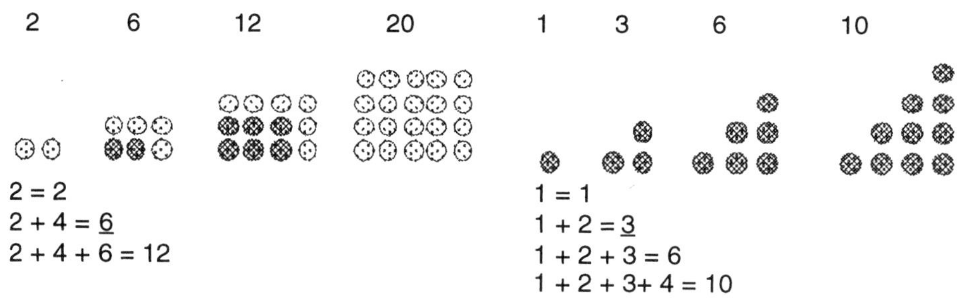

Figure 103. Black board image with patterns of rectangular and triangular numbers.

Rabea and Dennis noticed a connection: Each triangular number is half of a rectangular number. To illustrate this, several chips in the rectangular fields were turned (flipped) over to show a different color. According to the addition tasks corresponding to the triangular numbers, those corresponding to the rectangular numbers were to be found. In the following scene, these sums were to be shown by means of correspondingly colored chip configurations.

The short episode analyzed here is focused on the following particular justification question: Why is the chip configuration for the third rectangle produced by the student Nushin by flipping over the chips onto their red side correct? First the student Dennis formulates his explanation and after that, the teacher presents her justification of the correctness of the third configuration in the sequence of rectangular numbers.

153 T ...Dennis.

154 D Well the, the [*T points at the third rectangle*], the six, they are moved over this, too.

155 T Exactly. Actually, one can imagine that these six already go over there again [*points at the second rectangle and moves her hand over to the red chips of the third rectangle*], they stay there. And what happens in addition to this?

156 D Here, ehm, that is almost like this [*goes to the black board*], well, one only has to do that twice. Here [*points at the red chips of the second rectangle with his left hand*], there must be here, only some chips and here [*points successively below the two red chips of the second rectangle with the right index finger*]. That would then also be the same. That only ought to be double. [*points at the whole second rectangle with his right hand and looks at the chips of the third rectangular number in between*]

157 T If it was double, then it would be, eh,

158 D If there were two here now [*points below the two red chips of the second rectangular number*] and here comes another one now [*points at the spot below the blue chip on the right column of blue chips of the second rectangle*]. And here still come ... [*points at the left of the first column of chips of the second rectangle*]

159 T Yes, well. What have we done? You just said that we moved the whole, the six [*points at the second rectangle*] over here. [*points at the red chips of the third rectangle*] We have done that, too. And that‘s why they are red now [*circles the red chips of the third rectangle with her hand*]. Because that‘s that which we already have here. Thus that is here again [*circles „2 + 4“ in the sums of the second and third rectangle with chalk*]. Well, and what are these out here? The blue ones, what are they doing there? [*points at the hook formed by the blue chips of the third rectangle, Dennis goes back to his seat*] Bastian.

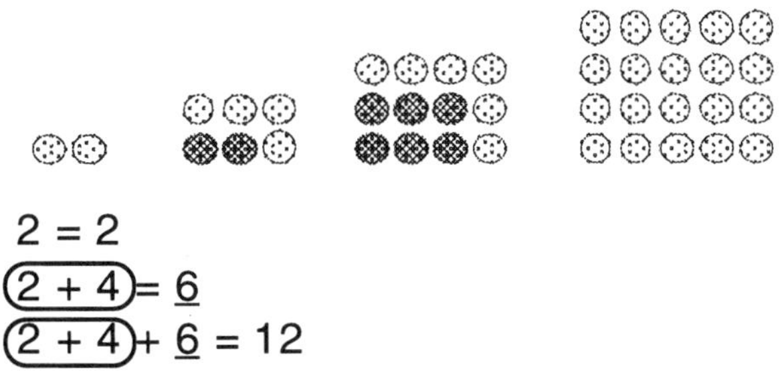

Figure 104. The teacher circles sums.

160 B These are the ones which remained first, being those which are newly added.

161 T Exactly, they are these six [*points at the hook formed by the blue chips of the third rectangle*] plus six [*points at the term „6" in the sum of the third rectangular number and underlines it with chalk*], which we have here. Then we arrive at twelve again. [*points at the sum „12" of the third rectangular number*]

This instruction scene with two attempts of justification can be summarized as a structured sequence of smaller interaction phases.

Phase 1 (153 – 158) Dennis develops a justification how the third pattern could develop out of the second one

The teacher draws students' attention to the 6 red chips in the third rectangular pattern (Figure 103). Dennis' first attempt at explaining is: „Well the, the [*T points at the third rectangle*], the six, they are moved over this, too." (154). [Presumption: Dennis means that the six red chips of the third configuration are moved over the two red chips of the second configuration, while the teacher interprets this in the reverse way: The 6 chips of the second configuration are moved over the red chips of the third configuration.] The teacher reacts to Dennis' statement as if agreeing with it, but is in fact proposing a divergent interpretation: „Exactly. Actually, one can imagine that these six already go over there again [*points at the second rectangle and moves her hand over to the red chips of the third rectangle*], they stay there."(155).

Dennis goes on with his interpretation. He explains how one can obtain the third rectangular configuration from the second one (not how one can identify the second configuration within the third one, which is probably what the teacher has in mind): „Here, ehm, that is almost like this [*goes to the black board*], well, one only has to do that twice. Here [*points at the red chips of the second rectangle with his left hand*], there must be here, only some chips and here [*points successively below the two red chips of the second rectangle with the right index finger*]. That would then also be the same. That only ought to be double. [*points at the whole second rectangle with his right hand and looks at the chips of the third rectangular number in between*]" (156). With this statement, Dennis tries to explain how the two present red chips can become four during the transfer.

Now it must still be made obvious how the quantity of blue chips changes. Dennis says: „If there were two here now [*points below the two red chips of the second rectangular number*] and here comes another one now [*points at the spot below the blue chip on the right column of blue chips of the second rectangle*]. And here still come ... [*points at the left of the first column of chips of the second rectangle*]" (158). This means that there must be one blue chip added, each at the top left and at the bottom right, and that he probably also finally refers to two further red chips.

Phase 2 (159 – 161) The teacher develops a justification different from Dennis' justification

The teacher gives a divergent interpretation. She refers to the point of view that the second dot configuration appears within the third configuration in a changed way. At this point she develops another argument. In the first step, she repeats the conception that the second dot configuration is „moved" onto the third one: „Yes, well. What have we done? You just said that we moved the whole, the six [*points at the second rectangle*] over here. [*points at the red chips of the third rectangle*] We have done that, too." (159).

The second dot field is colored red in the process of moving onto the third pattern: „And that's why they are red now [*circles the red chips of the third rectangle with her hand*]." (159). These red chips in the third pattern represent the old sum 2 + 4. In order to point this out, the teacher circles the term „2 + 4" each in the sums for the second and for the third dot configuration.

It is still necessary to explain what the blue chips in the third pattern mean from this perspective. Bastian says: „These are the ones which remained first, being those which are newly added." (160). Two perspectives are contained in this statement. On the one hand the view that has seemingly also been held by Dennis, one must construct the third configuration from the second one. And on the other hand the view held by the teacher, one is supposed to „recognize" the second configuration within the third one. „They remained first" means that these blue chips originate from the blue dots of the second configuration. And: „They are newly added" means that there is already an „old" part of the second pattern within the third pattern, but that new chips (here a blue hook) are added.

The teacher confirms Bastian's statement: „Exactly, they are these six [*points at the hook formed by the blue chips of the third rectangle*] plus six [*points at the term „6" in the sum of the third rectangular number and underlines it with chalk*], which we have here. Then we arrive at twelve again. [*points at the sum „12" of the third rectangular number*]" (161). The image displayed on the board looks as in Fig. 104.

In the following, the explanations made by Dennis and the teacher are again modeled as a sequence of signifiers. Dennis points at the second rectangular chip configuration and speaks to this. The whole diagram on the board (Fig. 103) and especially the cutting in the second and third geometric configuration of the rectangular numbers represents the reference context employed by Dennis and also by the teacher. The sequence of signifiers which Dennis communicates can be characterized as follows (Fig. 105).

In the first signifier, Dennis expresses the fact that the second configuration, especially the red chips, ought to be doubled. This idea is also repeated in the second signifier. With the help of the following signifiers, he describes in more detail which chips have to be added to the second configuration in order to obtain the third one. He points at the positions in the second configuration where red chips ought to be inserted (third signifier) and where blue chips ought to be added (fourth signifier) in order to obtain the third configuration in such a way. With the fifth signifier, Dennis suggests that more chips have to be added on the left side. Dennis does not form a new configuration – after all, it already exists – he imagines it and looks at the existing third configuration in order to do so. Dennis gives a construction procedure of the third configuration from the second one in which new red and blue chips are added without changing the pattern of the second configuration eg. by means of turning the chips over.

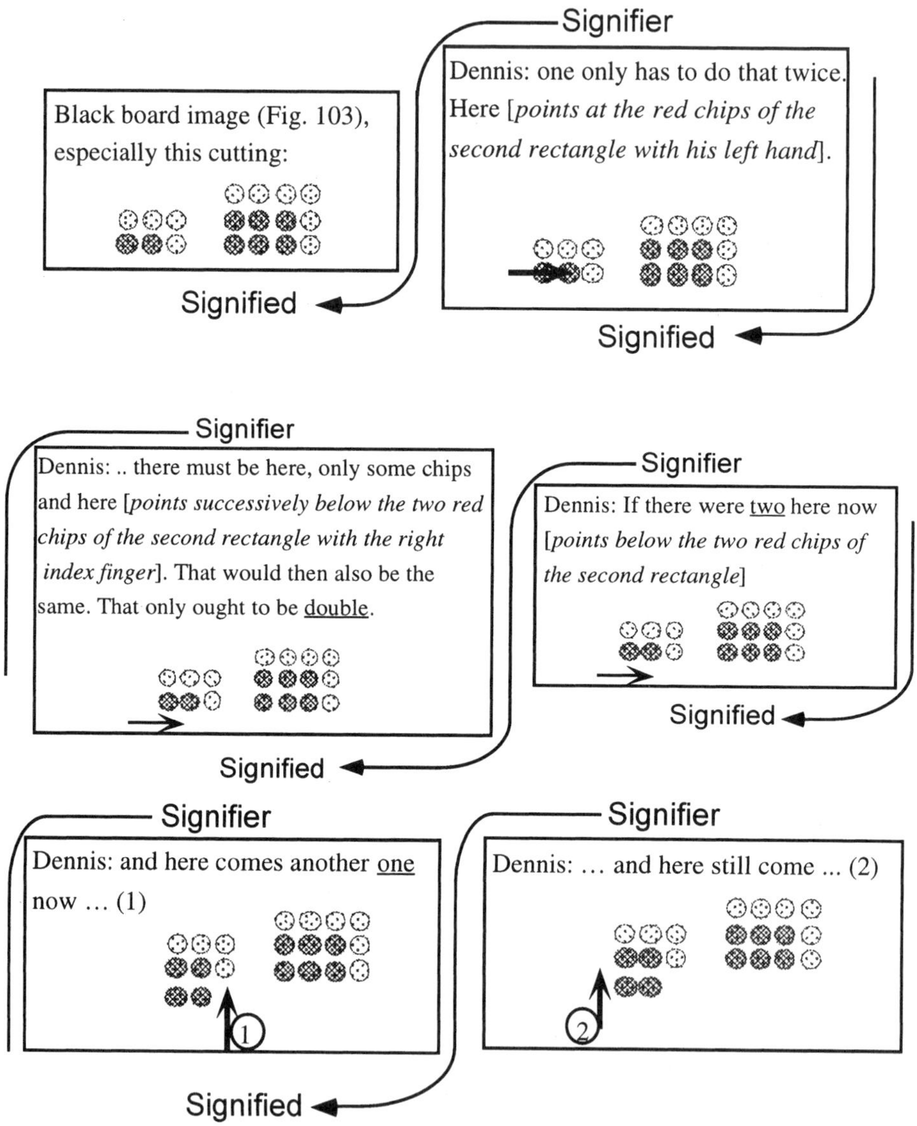

Figure 105. Dennis justifies the continuation of the pattern with a new rectangular configuration in a chain of signifiers.

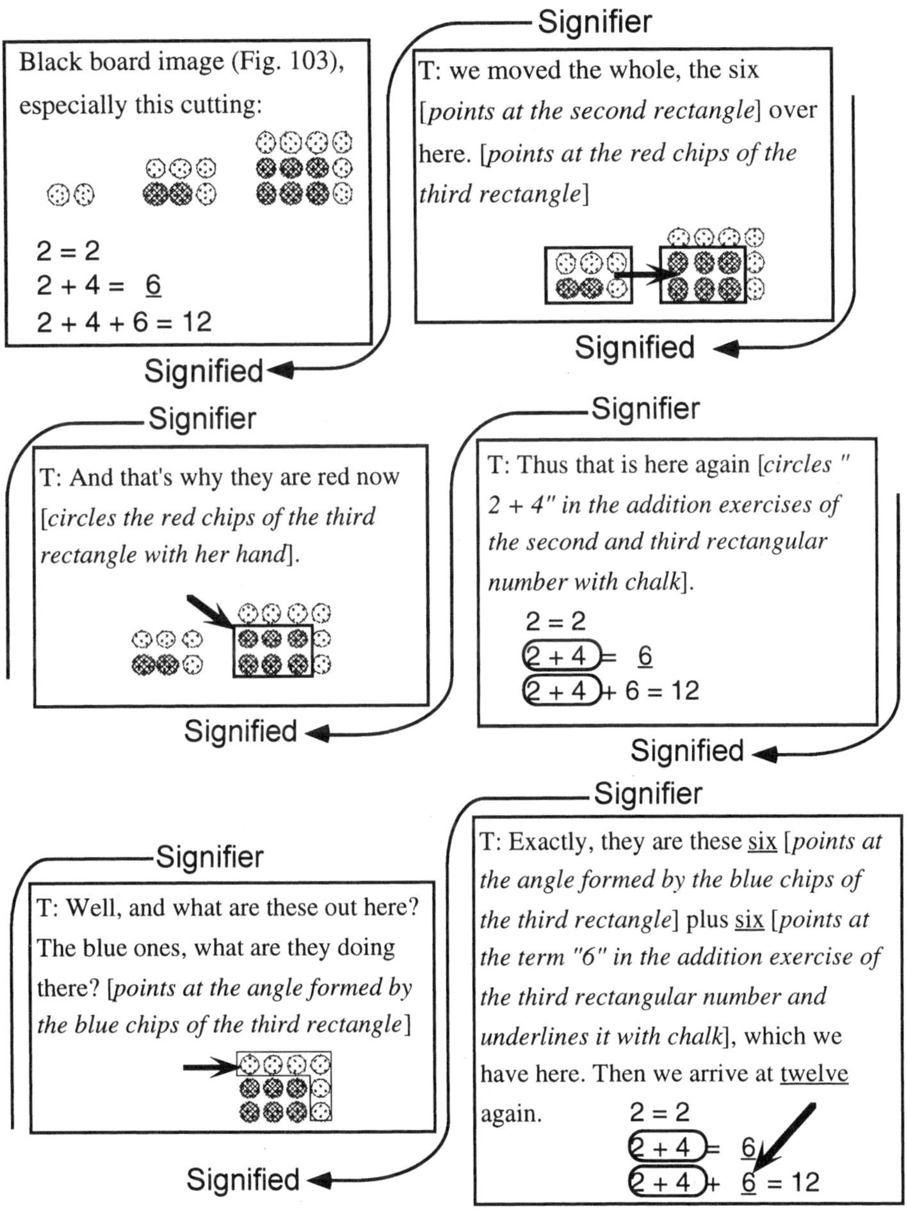

Figure 106. The teacher justifies the continuation of the pattern with new rectangular configuration in a chain of signifiers.

In the first part of her chain of signifiers (see Fig. 106; first and second signifier) the teacher suggests how the second configuration can be understood as a part of the third configuration by means of „moving it over" to the rectangle of the red chips in the third configuration (and simultaneous flipping over of the blue chips). With the third signifier she establishes a connection between the geometrical configurations and the arithmetical sums. The sum 2 + 4 for the second configuration can also be found in the sum for the third pattern 2 + 4 + 6. In order to point this out she circles the first two terms of the sum. In the fourth signifier the geometric hook of the newly arriving blue chips is emphasized and then the fifth signifier establishes a relation with the third signifier as the new term of the sum „plus 6".

How could these two justifications be judged from an epistemological perspective? As a justification for the fact why the third rectangular configuration is correct, Dennis describes a construction of the third configuration from the second one. The representation of this construction remains restricted to the concrete case. It could be – in principle – generalized, but this intention cannot be read from Dennis' description. Using the epistemological triangle (Fig. 107), one can state a *direct* connection between the new signs constructed by Dennis and the present context of the existing third configuration. In the reference context of the present second geometric configuration, additional new red and blue chips are to be put in the positions at which Dennis directly points with his fingers in the course of presenting his argument. In this way, he completes the second configuration to the third one; no relations or structures are used in order to do this.

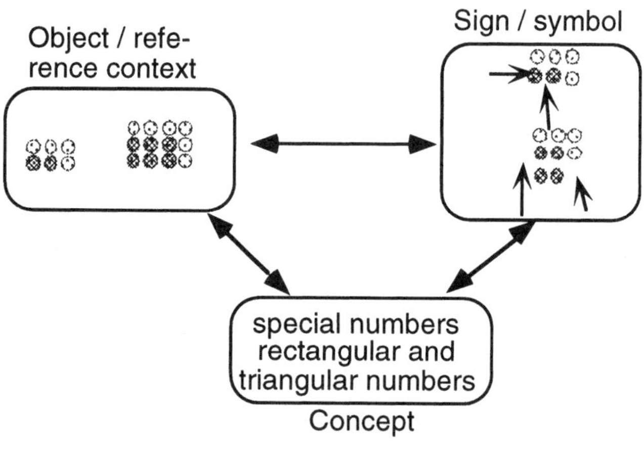

Figure 107. The epistemological triangle: Dennis completes the second configuration to the third one.

Dennis' intention is to prove concretely that the third configuration is correct. He seems to consider his construction procedure as a basis for the justification of the correctness of the third configuration. However, one could develop other patterns than the one established here by means of adding further red and blue chips to the given configurations. The correctness cannot depend on an ad hoc construction procedure, it must have something to do with the „form" that is to be continued.

Furthermore the correct chip pattern is not a fact which can be „logically" established. The mediation between the new signs, i.e. the third configuration and the reference context is of an empirical nature. The missing, concrete chips have to be added. In order to do this, no (geometrical or arithmetical) structures in the problem context are used, nor is this knowledge construction based on producing knowledge relations about the triangular or rectangular numbers. Thus Dennis does not construct an actual mathematical sign in this scene.

From an epistemological point of view, the teacher's knowledge construction can be seen as a justification in two parts. In the first part of the argument (159, 160), which refers mainly to the geometrical configuration, the teacher suggests how the second configuration can be understood as a part of the third configuration by means of „moving it over" onto the rectangle of red chips in the third configuration (simultaneously turning over the blue chips). The hook of blue chips in the third configuration consists of chips which are newly added. The uniqueness of the procedure of transfer from the second to the third configuration serves as a justification of the correctness of the third pattern. In principle, this procedure can be generalized. But the generality of the continuation principle is not directly spoken of – only indirectly, eg. in the remarks: „we moved them over here." (159) and „These are the ones ... being those which are newly added." (160).

In the second part of the argument (161), the teacher establishes a connection between the sums and the second and third chip configurations. The two terms of the second sum (2 + 4) and the first two terms of the third sum (2 + 4) represent the second configuration and the red chips in the third configuration, respectively: The hook of the blue chips is represented by means of the third term of the third sum, the „+ 6". This second part of the argument expresses a certain openness, as the relation between the terms of the sums (between the second and third pattern) is related itself to a relation between the two configurations (the principle of their construction).

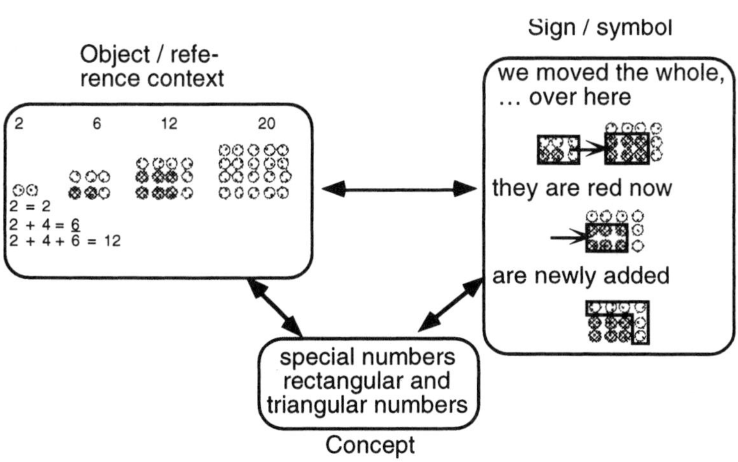

Figure 108. The epistemological triangle: The geometric continuation principle.

The epistemological interpretation in the triangle (Fig. 108) results in the following representation of the first part of the argument. In the reference context of the chip configurations the teacher produces a new sign by means of explicating a way of going from the second configuration to the third one. On the one hand, this sign (built by three components) is unequivocal and fixed for the concrete case; it is used to justify the correctness of the third configuration. Yet it could also be interpreted as a „general“ sign if one reads into it the principle of the transition to the next sized configuration. The interaction in this episode bears no directly recognizable references to the fact that this sign is supposed to or able to be transferred to other cases – neither by the students nor by the teacher. The second part of teacher‘s argument can be summarized in the epistemological triangle as follows (Fig. 109):

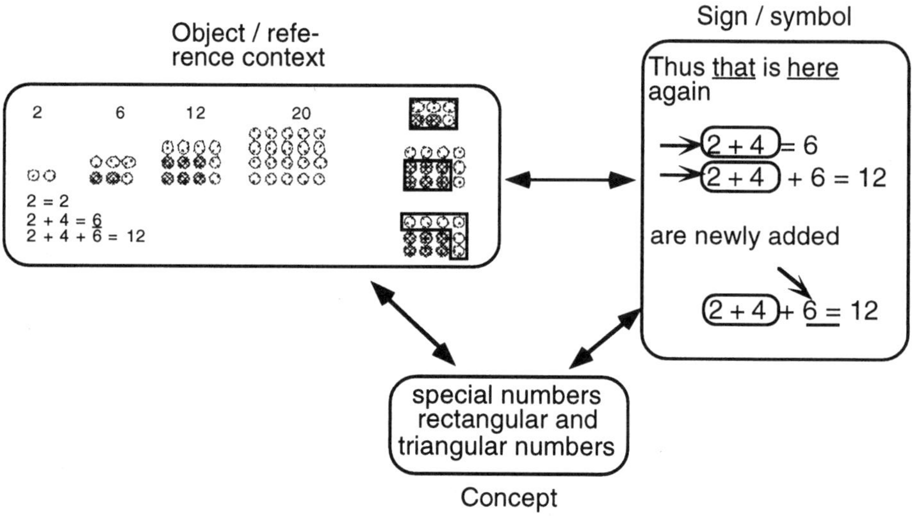

Figure 109. The epistemological triangle: The arithmetical continuation principle.

The teacher constructs new signs in the arithmetical structure of the sums by circling and underlining. On the one hand these signs are directly related to chip configurations: (2 + 4) is related to the second configuration and to the red chips in the third configuration, while the underlined 6 is related to the hook of blue chips in the third configuration. On the other hand, these signs are related to each other: they appear at the same time in the second and in the third sum, and they refer also to the transition from the second to the third configuration. The signs introduced by the teacher in the second part of her argument form a basis for structuring the arithmetical context of the sums, which not only gives the signs the function of a direct reference to elements of the reference context, but also provides them with independent, constructive aspects. With a continuation of this structuring more and more new relations could be built within this problem field.

The extensive analysis of the teacher‘s knowledge construction which has been carried out here shows in an exemplary way how the continuation principle for rectangular numbers could look in a first approach from a geometrical and arithmetical perspective. Even though the teacher refers only to the concrete case of the transition from the second to the third configuration, the study of this particular case could also be seen from a general perspective (whether this indeed happens in this scene or not cannot be supported by any of the participants‘ contributions). Thus the two mediations constructed by the teacher between the new signs (third configuration and arithmetical connections) are indeed interpretable as real mathematical signs, since the mediations refer to construction relations of the figurate numbers in the sense of a continuation and to structures which are embodied in the configurations. However, whether or not this is how they are indeed conceived in the interaction cannot be reconstructed in the analysis based on the observed statements of the participants.

4. EPISTEMOLOGICAL CASE STUDIES OF INSTRUCTIONAL CONSTRUCTIONS OF MATHEMATICAL KNOWLEDGE – CLOSING REMARKS

Qualitative analyses of mathematical classroom scenes represent attempts at better understanding human interaction processes about mathematical knowledge by means of a theoretical analysis. This contains a certain dilemma. On the one hand, the interactive events are supposed to remain uninfluenced by the observation and the chosen method of analysis and to „speak for themselves“ as far as possible. Yet, on the other hand, without a conscious choice of one‘s theoretical method and point of view one would remain on the surface of events and it would be very difficult to obtain deeper-lying conditions and explanatory descriptions of the observed phenomena.

In the epistemology-oriented qualitative analyses of the scenes from elementary mathematics teaching carried out here, an attempt was made to take this dilemma into consideration in the following way. On the one hand, an extensively justified method of analysis, elaborated in great detail with three methodological steps, has been used, which has also aimed at leaving the breadth and variety of the individual instruction scenes as much room as possible for their own representation. From this point of view, the epistemology-oriented analyses which have been presented in the previous sections represent very detailed and extensive case studies (of about 7 pages each) and these cases „speak for themselves“ extensively and thus make an understanding possible. This minuteness of detail of the single cases is necessary in order to be able to bring to bear the influence of the complexity and variety in the mathematical interactions.

At the same time, a point of view limited to particular epistemological interests has been chosen with the epistemological and communicative theoretical analyses of the instruction interactions. This interest had particularly in view the type of the observable social constructions of mathematical knowledge. Epistemological aspects and conditions of mathematical knowledge are not only understood as epistemological features and specific details of the long socio-historical development of mathematics, but the everyday mathematical communications and instructional interactions are also essential environments in which epistemological conditions, features and particular items of mathematical knowledge are constructed. Against this background, attention was paid among other things to the problem of whether and in which way „true“ mathematical signs and knowledge constructions can be realized within instructional interaction.

The epistemology-oriented analyses have confirmed the familiar point of view that elementary school students are not able to construct new mathematical knowledge and the necessary generalizing justifications with the „classical“ concepts of elementary algebra in order to describe the yet unfamiliar knowledge or to operate with it. In elementary school, the new mathematical knowledge is bound in a characteristic way to the situated learning and experience contexts of the students. With their attempts of developing and generalizing mathematical relations, the children are able to construct new knowledge and true mathematical signs with the help of their own situated descriptions. And in this way, they succeed in seeing the general in the particular and in naming it with their own words.

The new knowledge relations and mathematical signs which have been constructed by the students in their own way are not guaranteed by a new technical notation in which the old, concrete signs are substituted by new, abstract names. It is important that the essential conceptual relations are referred to in the construction and justification of the new knowledge, and this is already possible in the present frame of mathematical structures and designations. The epistemological analyses show that the *construction of a new mathematical relation* in the process of developing new mathematical knowledge precedes the mere *naming by means of signs*, i.e. a number sign becomes an abstract algebraic sign (a letter or a variable). The new knowledge is not a new name for an already known, old element of knowledge. The new knowledge develops on the present level of knowledge by means of the identification of a relevant, new conceptual relation, which possibly will only obtain a new designation afterwards.

In the next chapter I shall attempt to reach beyond the variety, independence and specifics of the detailed case studies presented in this chapter, and propose a classification of the interactive knowledge constructions. This will be done on the basis of detailed epistemology-oriented analyses, in which a justified qualitative comparison in the form of a further concentration on the important dimensions of the interactively-constructed knowledge is supposed to become possible. The central questions of this comparison will be, which particular items occur in mathematical instructional communications with young, learning students, which fundamental characteristics become effective in mathematical instructional conversations, which forms and developments towards „true“ mathematical interactions are observable and in which new mathematical knowledge is actually constructed.

OVERVIEW OF THE FOURTH CHAPTER

Following the extensive qualitative analyses of the selected teaching episodes with their interesting results regarding the variety and the particularities of the children's knowledge justifications and constructions, which have been elaborated in the third chapter, the fourth chapter tries to sum up and to unify this variety in a specific way.

Looking back on the previous chapters, the double role of mathematical signs / symbols as a connecting element between knowledge and communication is deepened one more time in chapter 4 (section 4.1). An essential aspect of mathematical signs and symbols consists in the type of mediation between the signs and respective reference contexts. Important characteristics in this are: (1) The signs as well as the reference contexts are not names or are not real objects, but embodiments of structures and relations. And (2), the referential mediation between sign and reference contexts requires the consideration of an assumed, relatively independent mathematical concept idea, which influences the mediation between sign and reference context. This implies a productive reciprocal relationship: The referential mediation is steered by conceptual mathematical knowledge and at the same time, conceptual mathematical knowledge emerges in the referential mediation.

The double role that the signs and symbols play for the knowledge and at the same time for the communication again refers to the context of mathematics teaching culture, in which the students are introduced to the use and meaning of the symbols by means of social participation. The epistemological necessities of mathematical knowledge and the conditions of communication can lead to different ways of interactive interpretating of mathematical signs and symbols in the culture of mathematics teaching (cf. the analyses in section 3.3).

In the early stage of instruction and learning processes, in which the students are slowly introduced to mathematical thinking and arguing and to the construction of mathematical knowledge, the range of the interactively constituted forms of knowledge interpretation goes from an empirically concrete to a relational structural understanding of mathematical signs and symbols (section 4.1).

A confrontation of communication, construction and justification of mathematical knowledge, as it occurs in the professional practice of researching mathematicians, with the teaching practice, in which the young learning students communicate about mathematics, is supposed to make the specific types of mathematical knowledge constructions clear (section 4.2). While the communication and justification of mathematical knowledge in the professional mathematical practice is based on a uniform, universal view of mathematics, according to which the ideal mathematical research objects are defined implicitly by means of conceptual relations (e. g. axioms), the objects or reference contexts of mathematical knowledge and mathematical signs in teaching and learning are more varied and non-uniform.

The initial question for this problem can be formulated in the following way: What is the relation of the referential mediation between reference context and sign to the role of the conceptual mathematical structures? The following two basic types can be identified: In the professional research practice, basic, consistent conceptual structures (axioms) serve for describing and then examining ideal mathematical objects in an indirect way. In elementary mathematics teaching, given objects, phenomena and patterns, which can be viewed, serve in the opposite way for deriving structural conceptual relations out of the perceivable features and thus constructing conceptual mathematical knowledge. For the knowledge construction the mathematical learning process first requires given objects and phenomena, for which examination regarding their structures leads to conceptual mathematical knowledge. In the professional mathematical practice the ideal object to be examined is defined implicitly by means of conceptual structural relations in a universal way, and then this object is further differentiated and elaborated in the examination of the derivable relations.

These differences between the professional and the mathematical communication in the classroom are important, as they make it clear that mathematical communication can not proceed immediately from one universal, uniform mathematical object, but that the students' mathematical knowledge must develop over several stages. This development contains different kinds of knowledge justification and construction. In section 4.2, four kinds of knowledge construction are distinguished, in which different justification contexts for the mathematical knowledge to be constructed are used. The kinds of justification vary between empirical justifications on the basis of given objects and features and justifications on the basis of consistent symbolic systems (e.g. axioms) (see section 4.2).

The problem of the development of mathematical communication in the classroom, which emerges over stages and different kinds of justification, is further discussed in section 4.2 in the frame of open versus closed communication and looking at the relation from old to new knowledge. Direct or unequivocal communication between teacher and students is accompanied by justification contexts, in which given (concrete) objects with perceivable (empirical) features are used. Open communications, on the other hand, make potential justifications in the communication possible, which are based on relations and structures. Closed communications with empirically oriented justifications usually have the old, familiar factual knowledge in view, whereas open communications aim to work towards the construction of possible new relations on the basis of mathematical structures.

The many-layered justification contexts of mathematical argumentations and knowledge constructions in interactive teaching-learning-processes, as they have been analyzed in section 3.3, are classified with the help of the two-dimensional analysis grid in section 4.4. This classification does represent a reduction regarding the extensiveness and variety of the information, which resulted from the three-step, epistemologically oriented analyses. On the other hand, it makes a comparison possible between the different interaction types in the selected episodes. This comparison takes into consideration in an ideal way the two poles of „empirical situatedness" to „relational generality" and of „imparting of factual knowledge" to „construction of interpretations", as they are taken up again more closely in sections 4.1 and 4.2.

Altogether, ten interaction episodes are classified with the grid. There are classifications of interactions, in which the knowledge constructions are based on empirical circumstances and factual knowledge. Besides, such interaction scenes can be identified, in which arbitrary relations are constructed and interpreted in a relatively loose and free way – in the sense of a schematic or exaggerated generalization. Such mathematical interaction episodes, which have been classified as a balance between the extreme poles along the two dimensions of the grid have been evaluated as successful, true mathematical communications and constructions. This balance allows expression for the communication of the connection between the present knowledge and the new, relational knowledge to be constructed, and it keeps up the necessary tension between the empirical situatedness and the intended relational generality in the epistemological dimension.

Following this division into three essential different types of knowledge construction, as they result from the classification with the analysis grid, three forms of interactive mathematical knowledge constructions are described in section 4.4: knowledge construction as (1) continuation of familiar, situated factual knowledge, (2) a balance between consistent base knowledge and new knowledge relations, and (3) introduction of isolated knowledge structures.

CHAPTER 4

EPISTEMOLOGICAL AND COMMUNICATIONAL CONDITIONS OF INTERACTIVE MATHEMATICAL KNOWLEDGE CONSTRUCTIONS

1. SIGNS AS CONNECTING ELEMENTS BETWEEN MATHEMATICAL KNOWLEDGE AND MATHEMATICAL COMMUNICATION

The previous considerations and analyses made it obvious that the mediation between signs and reference contexts is subject to particular conditions, when it is about mathematical signs, which try to take hold of the specific character of mathematical knowledge. First, the kind of mediation between mathematical signs and reference objects depends on the activities of the individual.

> … a sign always remains framed by the practical activity of the individuals and to conceive the sign as a semiotic object functioning in a map or environment where the specific characteristics of the activity has to be taken into account (Radford, 2003, p. 50).

Further, this mediation between mathematical signs and reference objects depends on the epistemological conditions of the theoretical mathematical knowledge. The attribute „mathematical" in „*mathematical* sign" or „*mathematical* symbol" is not to be reduced to mere enumeration of particular forms of notation, ways of writing and chains of symbols (as claimed, for example, in Davis & Hersh, 1981, p. 122 ff.). Mathematical signs are first of all also very general signs, which are indispensable means for human communication and human thinking. As with all signs, mathematical signs are also carriers of knowledge and elements of communication with others.

Classroom communication, like everyday communication, uses natural language and the possibility of direct pointing to the object of conversation. But in instruction, these means of communication serve, first of all, the purpose of exchanging and recording mathematical ideas. Descriptions in ordinary language, designations by means of specific names, direct pointing to and referring to something, are all preliminary forms of interactively produced signs for coding and developing aspects of mathematical knowledge. Furthermore, classroom communication uses more subject matter specific, mathematical signs such as technical expressions, written notations, mathematical coding signs, variables, special mathematical symbols, diagrams etc.

Forms of mathematical signs used in social interaction range from everyday ways of speaking and direct pointing to the standard forms, which have long been established in the mathematical discipline. In the mathematics classroom culture, learners are supposed to become familiar with different forms of mathematical signs used in the interaction, and acquire their use by means of social participation, rather than use given formal signs according to strict rules. For the learners, signs, their interpretation and generalized used emerge only gradually. One cannot furnish the learner with mathematical signs in their essential meaning at the beginning of the learning process.

In everyday mathematics instruction, multiple forms of mathematical signs can be observed:

- verbal formulations, own words with descriptions based on concrete examples
- communication by means of pointing and referring to (deictic)
- mathematical symbols: number signs, operation signs, letters, variables, ...
- arithmetical sentences such as sums or products, and equations, systems of equations, ...
- tables, geometrical diagrams, graphs of functions, ...

These forms of signs range from the spontaneously and interactively created signs that are situated in particular learning environments to sign formats that are conventionalized and commonly used in mathematical instruction. In addition to the „evolving interpreting games" (Sáenz-Ludlow, 2001, p. 20ff), the epistemological dimension of mathematical knowledge requires taking into account the epistemological conditions of knowledge as factors that are relatively independent from the communicating subjects. The epistemological conditions play an important role in the mediation between the sign and the reference context.

> [I]n epistemology as well as in logic or communication theory we have to acknowledge that communication and knowledge are possible only when there is some Other, which is neither fused with the subject nor is totally different from it (Otte, 2001, p. 5).

In the previous chapters, the particularity of a sign as a *mathematical* sign was characterized by two essential aspects. (1) The sign as well as its accompanying reference context are themselves embodiments of something else, namely of structures and relations. (2) The mediation between sign and reference context requires the use of pre-existing, relatively independent mathematical concept relations. In active mathematical learning processes, such mediations between sign and reference context are carried out by the participating subjects in direct interaction. The question arises then how such interaction can take into account the epistemological conditions of mathematical knowledge. The generality of the socially produced mathematical signs depends, among other things, on whether this takes place in a direct, face-to-face communication, or in a communication with non-present partners (cf. Radford, 2001). In temporally longer, socio-historical processes of developing mathematical knowledge, more and more conventionalized, fixed and compulsory forms of mathematical signs develop by means of mediations between signs and reference contexts.

An important problem is to understand how mathematical signs can be already produced in elementary, initial mathematical learning processes, and if this is at all possible. An important central assumption is that neither the reference contexts nor the signs themselves are equated with mathematical knowledge, i.e. with conceptual relations. On the one hand, this problem refers to the difficulty of „grasping" mathematical knowledge. On the other, this problem contains the possibility that conceptual knowledge can be grasped independently to a certain extent from the *form* of the signs. One must see that signs are obviously unavoidable and indispensable in order to record mathematical knowledge, yet the type or the form of the signs is not established. In order to record algebraic relations, the typical algebraic signs, thus letters and operation signs, are not necessarily required, yet these signs themselves are not the algebraic knowledge in question. Up to a certain degree, one is free to choose other sign forms, and this possibility of choice of signs allows for the change, development and optimization of mathematical signs for a suitable characterization of the invisible conceptual mathematical knowledge.

The interaction processes, observed in the episodes, were situated in concrete learning environments based on specific examples of arithmetical or geometrical connections and contexts. Children have interactively constructed their mathematical signs within these given learning environments using their own designations and referring to the situated features. These young students could not simply take pre-given, general or suitable signs – letters, for instance –, with which the general structures in the arithmetical problem fields could be produced and justified.

The double role of signs and symbols – as carriers of the knowledge in question and elements exchanged in communication with others – refers to the particular complementary relation between knowledge and communication. From this point of view, the development of knowledge itself as well as its appropriation by the learner through communication is a social and cultural process, which is not completely and exclusively determined by objective, explicitly given conditions. Signs and symbols are central components of culture in general (see chapter 1). In this regard, one can understand cultural communication as social transmission of cultural knowledge. The use of the cultural signs and symbols practised within the respective culture determines the cultural values and the relevant cultural knowledge, and inversely, the present cultural knowledge also shapes the respective culture.

If one understands everyday mathematics instruction as a particular culture, whose aim is the introduction of (school-) mathematical knowledge into the classroom culture, then the signs and symbols used here again play an important role in the interrelationship between the culture of mathematics instruction and the cultural knowledge considered essential here, namely school-mathematics. Mathematical interactions in the frame of the mathematical classroom culture can be conceived of as processes of social mediation of mathematical understanding and of (school-) mathematical knowledge. And conversely, conceptions of the nature of (school-) mathematical knowledge shape the features of the mathematical classroom culture.

This perspective on mathematics teaching as a particular culture, in which the interrelationship between (school-) mathematics and the cultural aspects of teaching and learning is considered, has been described for instance by Stigler and Hiebert (1999) in their book „The Teaching Gap“. In particular, these authors thematize the particular nature of mathematical knowledge as a cause of different instruction patterns (Stigler & Hiebert, 1999, p. 89). The question of whether mathematics is seen as a product or as a process – to quote Freudenthal – is essential in the choice of instructional activities and for the children's learning. An important aspect of Stigler and Hiebert's considerations is the idea of instruction as a cultural activity. This means that instructional activity cannot be conceived or even simply be learned as a collection of different, elaborated and well-defined techniques. Like other cultural activities, the activity of instruction cannot be completely acquired by means of fixed rules or strict „directions for use“. This activity is learned mainly through engaging in it, practising it and sharing one‘s experience with colleagues in a social context. Furthermore, Stigler and Hiebert emphasize that the activity of teaching represents a system:

> Teaching is a *system*. It is not a loose mixture of individual features thrown together by the teacher. ... This is a very different way of thinking about teaching. It means that individual features make sense only in terms of how they relate with others that surround them. It means that most individual features, by themselves, are not good or bad. Their value depends on how they connect with others and fit into the lesson (Stigler & Hiebert, 1999, p. 75).

In everyday teaching practice, the complementarity between the epistemological conditions of (school–) mathematical knowledge and the interactive conditions of mathematics instructional communication leads to different expressions of the character of the thus constituted mathematical knowledge. In this chapter, I will identify essential types of knowledge produced interactively in the analyzed episodes. I will use a grid of analysis in which the two dimensions, „epistemological conditions of mathematical knowledge" and „interaction conditions of mathematical communication" are related to each other (cf. section 2.3).

The central connecting links between (school-) mathematical knowledge and instructional communication are the signs and symbols (in the generality described before) that are used in the interactions. Thus the way signs were used and interpreted in the communication will be an essential indicator to decide what kind of knowledge form had been produced in the interaction. The interactively produced knowledge forms will range from empirical and concrete forms to relational and general ways of using and interpreting the signs and symbols. If, for instance, the (general) signs in the instructional communication had been used as names for direct designation of concrete things or features, then the interactively constructed mathematical knowledge form will be considered empirical. If general and relational ways of interpretation seemed to be intended, the knowledge form will be considered as relational.

A problem for the instructional development of mathematical knowledge in the interrelationship between epistemological and interactive conditions consists of the fact that, in spite of the necessity of obtaining general and relational mathematical knowledge, learning and understanding depends, to a certain extent, on contexts, situations and concrete objects and material. According to the epistemological conception presented here, context-dependence and situatedness of mathematical knowledge are not questioned (cf. Steinbring, 1998a). However, the crucial point to consider is the *kind* of the context-dependence or of the reference to concrete objects and features: Is the epistemological interpretation of the mathematical knowledge directly linked with the *finished*, empirical features of the situation. In other words: is knowledge derived from the concrete features, or does the situation with its structure serve as an *open* reference context, which must first of all be interpreted in a suitable way and which also allows for new interpretations? This point of view understands the situated character of mathematical knowledge as a relation between signs / symbols and reference contexts, in which the present context does not directly explain the knowledge, but the situated context can be interpreted in a different way and can be used as an embodiment of structural connections, which allow for the construction of new mathematical knowledge.

A central problem for the construction and justification of new mathematical knowledge consists of the fact that neither the (concrete, situated and structural) reference contexts nor the sign and symbol systems alone directly contain the mathematical concept. Signs and reference contexts form a possible basis for construction in so far as with their help, one can refer to structural connections and relations, or model structural connections interpretable as embodiments of concept relations. This means that mathematical signs and reference contexts do not reflect the new constructed knowledge directly and immediately, but they are used as indispensable, iconic carriers of knowledge in the sense of hints to other structural relations of the concept. Thus, when dealing with mathematical signs and accompanying reference contexts, students face the particular interpretation problem of always distancing themselves from the concreteness of the situation and of seeing, interpreting or discovering „something else", another structure within it. This problem is discussed in mathematics education literature under the key idea: „Seeing the general in the particular" (cf. Cobb, 1986; Mason & Pimm, 1984).

2. PARTICULARITIES OF MATHEMATICAL COMMUNICATION IN INSTRUCTIONAL PROCESSES OF KNOWLEDGE DEVELOPMENT

From a comprehensive theoretical perspective on mathematical knowledge and the epistemological conditions of its development, the main interest of this study is to gaining a better understanding of the question how, in mathematics teaching, students interactively construct mathematical knowledge in the frame of their means and possibilities. In the previous chapters, it has been repeatedly stressed that, from an epistemological point of view, mathematical knowledge consists essentially of „conceptual structures and relations". These conceptual structures and relations are subject to a difficult form of existence. On the one hand, they are independent to a certain extent, while, on the other hand, they require the „objects / reference contexts" and the „signs / symbols", as modeled by the epistemological triangle. How do the conceptual structures and relations develop in different practices of communication? Or, what interactively produced forms of mathematical knowledge can be characterized in different practices of communication? The interpretation of reference objects and signs carried out by the participants in the communication is essential in answering these questions. A purely phenomenological or empirical point of view leads to a different type of mathematical knowledge than a relational interpretation. Furthermore, it is important to consider the question of whether the reference contexts with their accompanying signs are used in the communication as *the* basis for deducing mathematical knowledge, or if in the communication other relations between the „conceptual structures and relations" and the „reference contexts with accompanying signs" are produced.

I will elaborate on this basic problem in more detail. In the epistemological triangle, this problem can be interpreted as one of the relations between the „conceptual structures" and the necessary „environment", which is produced within the communications and interactions, i.e. the „objects / reference contexts with accompanying signs / symbols" (see Fig. 110).

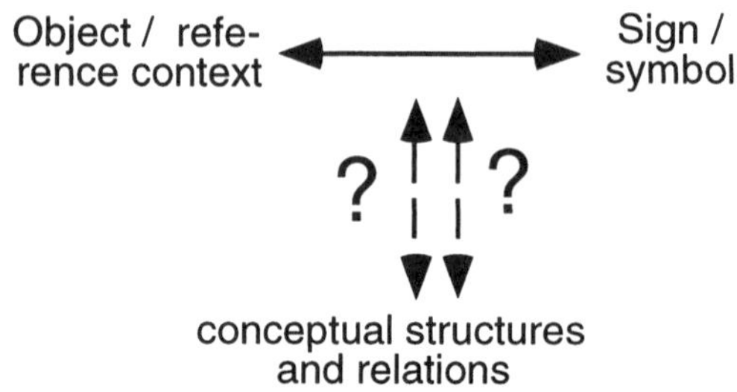

Figure 110. The epistemological triangle: The role of conceptual structures.

In the presentation of the „epistemological triangle" as an instrument of theoretical analysis, it has been emphasized that an interrelationship between the mediation between „objects / reference contexts" and „signs / symbols" on the one hand and the „conceptual structures" on the other hand is necessary for the construction of true mathematical symbols. On one side, the „conceptual structures" are a necessary, independent prerequisite for this mediation, and on the other side, new conceptual structures will develop from the mediations between „objects / reference contexts and signs / symbols" during the further, interactive development of the knowledge. This is a somewhat ideal, epistemological characterization of the relation between the „conceptual structure" and the „reference contexts with accompanying signs".

In everyday mathematical communication processes, different, interactively produced relations can be observed. There is also a fundamental difference between how children learning mathematics can interpret and use this relation and how this relation is organized and regulated by research mathematicians.

With regard to the difficult interrelationship between „objects / reference contexts and signs / symbols" and the „conceptual structures", three important problem perspectives, which have already been mentioned in chapters 1 and 2, will be taken up again at this point and then applied in the classification of the students' interactive knowledge constructions. The following three central problem perspectives are considered: (1) The *unequivocality* of professional mathematical argumentation and the *ambiguity* of students' mathematical argumentation; (2) *Open* vs. *closed* communication in instruction; (3) The dialectic relation between *old* and *new* mathematical knowledge.

In order to discuss these three problem dimensions, two different points of view towards the epistemological triangle (Fig. 110) will be taken. These are expressed in the following diagram (Fig. 111).

In this representation, the repeatedly discussed problem of the relation with the „conceptual structures" as „object" becomes visible again: Are the „conceptual structures" a result derived from the given „object", or do the „conceptual structures" serve as a basis for the construction of the idealized „object"?

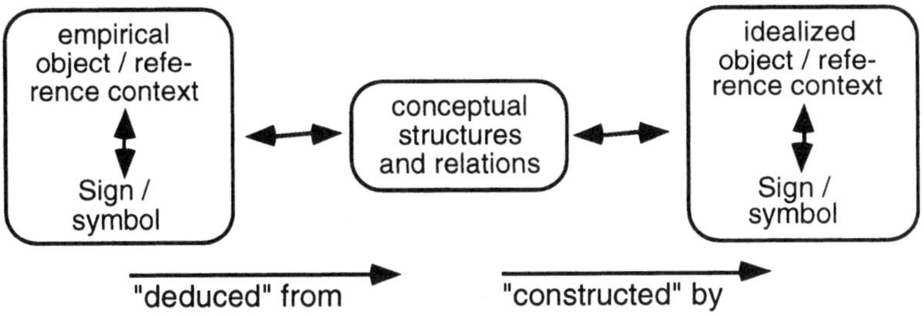

Figure 111. Different interpretations of „conceptual structures" within the epistemological triangle.

(1) The unequivocality *of professional mathematical argumentation and the* ambiguity *of students' mathematical argumentation*

In her extensive work about the unequivocality of professional mathematical communication and argumentation, Bettina Heintz (2000; see also chapters 1 and 2) has elaborated the modern mathematical proof as the crucial means for the increasing unequivocality of this professional communication. In her analysis, she stresses that a changed conception in the relation between the „objects" and the „conceptual structures" in the history of mathematics was essential for this development.

> In the course of the 19th century, the ‚naïve abstractionism' ... of earlier mathematics was overcome and replaced by objects, which were defined in an exclusively internal mathematical way. This ‚work on the concepts' had also become urgent because mathematicians increasingly used concepts, which could no longer be understood as idealizations or abstractions from empirical experiences, but which had an exclusively ‚fictional' character. ... In the process of this ‚theorization (Jahnke, 1990) or ‚de-ontologization' (Bekemeier, 1987, p. 220) of mathematics, concepts, which had been taken for granted before, were successively questioned and transferred into an explicit system. ...
>
> In the process of this conceptual reflection and reconstruction, essential parts of mathematics lost their natural and visual character. Those mathematicians who broke with tradition at this point of time, and who replaced the ‚natural' given by theoretical constructs, as it was often criticized by the orthodox side, were still clearly aware of the breach they committed (Heintz, 2000, p. 263, 264).

A strict and unquestionable basis of justification on which the unequivocality of mathematical argumentation could develop and in this way become a set of precise rules of communication between professional mathematicians was made possible only by leaving the given empirical objects and by constructing idealized mathematical objects by means of defining conceptual relations (Heintz, 2000, p. 221). Given objects and immediate observations are interpretable in many ways and can lead to contradictory conceptions.

> Even when arguments are deductively constructed and argue with the help of logical rules, when these arguments require a common knowledge and rely on intuition and visualization they are more at risk of dissent than a formal argumentation which one can hardly avoid even if it is contrary to intuition and experience (Heintz, 2000, p. 274).

In terms of Fig. 111, the position of the professional mathematical argumentation represented here can be characterized as follows: the objects used in mathematical argumentation no longer possess any immediately given visual features, but are contradiction-free constructions based on defining conceptual relations. Seen historically, this has been a fundamental changing process, in which visual circumstances were replaced by artificial objects. Even if ontological or idealized object aspects play no explicit role in official professional mathematical communication, i.e. in writing proofs for publication in professional journals (cf. Heintz, 2000, p. 221), one can still assume that, when working on a mathematical argument, the individual mathematician cannot do without such idealized object conceptions in his consciousness or private world of ideas (Heintz, 2000, p. 220ff.).

As a consequence it can be stated that in professional mathematical communication, the contradiction-free conceptual structures and relations dominate and determine the ideal objects which are not interpreted any further.

> In contrast to content related axiomatics, formal axiomatics give up the content related qualification of the axioms... Axioms are assumptions of a hypothetical kind, whose content related truth is not under discussion. Axioms are true if no contradiction results from them, and the same holds true for the existence of mathematical objects... (Heintz, 2000, p. 265).

Contrary to the professional mathematical communication, the emphasis in the mathematics classroom communication is on the relations between „conceptual structures“ and „object“. The aim of mathematical instruction is, among other things, to introduce children to a meaningful participation in the specific forms of mathematical communication. Things that are taken for granted by the experienced professional mathematician, namely the unproblematic use of proof as the accepted principle of communication, are a long-term goal for the student on their way to becoming a mathematically thinking person. Thus the kind of mathematical communication and the justification contexts accepted here are essential objects of learning for the student. Professional mathematical communication cannot be directly transferred into the instruction process. It is necessary to construct the independent conditions of developing instructional communication for students while taking into account the essential characteristics of mathematical communication.

As for the professional mathematician, the conceptions about the ideal object are of a private nature and limited to his individual consciousness. For the early learners, visual and content related conceptions about the object are an important first basis for the justification of conceptual structures and relations, and the students must reflect these conceptions jointly and together with their teacher in instructional communication. In terms of the diagram in Fig. 111, instructional mathematical communication will be characterized more by means of the relation between an (empirical) object and conceptual structures. In the next section, the question will be answered as to which forms of the relation between „object and conceptual structures“ can be observed in particular cases of real instructional interactions. A classification of the ten epistemologically analyzed teaching episodes (presented in Chapter 3) is a basis for an answer. One should not assume in a simplistic way that the natural, empirical circumstances only exist to produce the elementary conceptual structures. Even in elementary school mathematical communication, the conceptual relations are not mere abstractions from the given empirical and visual features but they are conceptual structures of their own.

The basic forms of relations between objects and conceptual structures, which occur in instructional conversations, and thus the respective contexts, which are used for the mathematical justification and argumentation, can be characterized in the following way.

- Empirical justification context: The conceivable features of the given (empirical) object represent the justification context for conceptual structures.
- Algorithmic justification context: Arithmetical (or partly algebraic) patterns, rules and laws represent the justification context for conceptual structures.
- Situation-bound relational justification context: Detachments of concrete features and giving new interpretations to features of the (empirical) objects lead to first intentional and constructed structures in the given situation and in this way become a justification context for conceptual structures.
- Independent relational justification context: Mathematical relations are coded independently with the help of symbolic systems and (partly) serve for the definition of the features of the ideal mathematical object and this becomes the justification context for conceptual structures.

The development of the above mentioned types of mathematical justification contexts is accompanied by a development in the interplay of relations between the two poles, „object“ and „conceptual structures“. While the „empirical and algorithmic justification“ represents cases of instructional mathematical communication and justification based mainly on the given features, patterns and concrete ways of proceeding, the „situation-bound and independent relational justification“ starts the interplay between the conceptual structures and the features of the ideal object to be constructed. This interplay must be clearly distinguished from the professional approach to theoretical constructions of artificial mathematical objects (for instance by means of axiomatization) (as it is described in Heintz, 2000). This interplay between the conceptual structures and the features of the ideal object is an important and genuine form of communication and justification, and it represents an indispensable pre-requisite for the development of professional mathematical argumentation.

The above four kinds of justification that can be found in instruction mark stages of development and present a challenge in helping students to become competent in using the „independent relational justification“ in their argumentation. Altogether, they can be understood as forms of a yet to be developed set of rules for students‘ mathematical argumentation similar to the set of conventional rules of communication among professional mathematicians.

In the classification carried out in the following section, these types of justification are used and further concretized. First of all, the reference to general instruction communication is to be recalled, because mathematical communication is always influenced by instructional communication (see section 2.2).

(2) Open *or* closed *communication within instruction*

General communicative structures and patterns in mathematics teaching influence the kind of interactive mathematical knowledge construction and argumentation. It was assumed, in particular, that the problem of the instructional interaction is determined by two factors: the content of instructional communication, namely mathematical knowledge, and the asymmetric structure of the instructional communication where the teacher who knows must „mediate" correct mathematical knowledge to the students who do not yet know it.

The way communication (in the sense of Luhmann, 1997) is started and guided by the interactive use of successive signifiers (see section 1.3 and section 3.3) is an essential aspect of the communicational analysis of mathematical instructional interactions carried out here. Ultimately, the interpretations of the communicated signifiers are open in the sense that the assignment of possible signifieds in the communication must be carried out independently by the receiver of the message and that the signified is not unequivocally pre-determined. At this point, the goal of teaching mathematics – namely mediating correct knowledge – can have an effect on the communication in so far as the participants in the instructional communication subliminally assume fixed, standard signifieds for the communicated signifiers, and by trying to reconstruct these signifieds they lose sight of the open, yet to be produced mathematical relations as the essential signifieds to be constructed (see section 2.2).

Now, this problem of mathematical instructional communications can be connected with the four justification contexts for mathematical argumentations by students. Teaching is expected and required to mediate correct mathematical knowledge to the learners. Therefore the participants in instructional communication increasingly concentrate on directly communicable and unequivocal contents. Instructional communication constantly tends to assign unequivocal signifieds – intended by the teacher – to the reciprocally communicated signifiers. These unequivocal communication contents can be found especially within empirical and algorithmic justification contexts. This leads to a reciprocal action, which mutually reinforces itself, between direct instructional forms of communication and the more empirical, visual and procedure-oriented justification contexts for mathematical knowledge.

True mathematical communication (cf. section 2.2) must leave room for openness and interpretability of the communicated (mathematical) signifieds. In the frame of mathematical communication, this condition is taken into account especially when the interpretation of „conceptual structures and relations" increasingly moves into the focus. And the „conceptual structures and relations" are not simply derived from natural circumstances, but they are applied as means of construction for new (ideal) mathematical object features. In this way, the signifiers communicated in the mathematical communication (thus the mathematical signs and symbols) obtain their necessary openness and their expected character for the interpretation by the persons participating in the mathematical communication. Thus a true, i.e. open and interpretable, mathematical instructional communication is in close connection with the situation-bound relational and the independent relational justification contexts for mathematical knowledge. At the same time, this represents an ambitious challenge for mathematics instruction and for active construction of mathematical knowledge, as traditional teaching communication is often opposed to open, intentional interpretation and active construction of mathematical relations.

The kind of instructional communication – direct or unequivocal and intentional or open forms of communication – corresponds to the justification contexts for children's mathematical argumentation in instruction. Furthermore, connections between these justification contexts and the conceptual pair „old and new mathematical knowledge“ can be produced.

(3) The dialectic relation between old *and* new *mathematical knowledge.*

The problematic relation between old and truly new mathematical knowledge was discussed in detail in section 2.1. The underlying core idea was expressed using the following quote from Max Miller.

> From a learning or developmental theory it can be ... legitimately expected that it gives an account of how the *new emerges* from the old in development. ... Any answer to this question ... is ... subject to the following validity criterion: one has to be able to show that the new in the development requires the old in the development and still systematically exceeds it, otherwise there can be no new or the new is already old, and then the concept of „learning“ or „development“ loses any sense (Miller, 1986, p. 18).

What is the connection of the problem of development of truly new mathematical knowledge to the four justification contexts for children's mathematical argumentation in mathematics instruction? Miller adds three important questions to the central validity criterion for truly new knowledge:

> How can the validity of the already acquired (old) knowledge ... be shaken or relativized ... for an individual? How can the individual gain new relevant learning experiences, which systematically transgress his present knowledge? And how ... can the individual feel compelled to further develop his knowledge? (Miller, 1986, p. 18/19).

On the one hand, these questions aim at the problem of an actually new knowledge, which is not already present in old knowledge (for instance as a schematic or logical conclusion). On the other hand, they aim at the social conditions under which learners are able to construct this ambitious new knowledge.

In „learning processes" that are organized exclusively according to the individualization principle, and restricted to the single learner and their cognition, the individual – according to Miller – can ultimately only add knowledge derived from their already present knowledge, which is therefore no new knowledge, but identical with the old knowledge. Only social processes make it possible to develop new knowledge by means of contrast, contradictions and new interpretations.

> Only in the social group and thanks to the social interaction processes between the members of a group can the individual gain experience that makes the fundamental learning steps possible (Miller, 1986, p. 20/21).

The relation of new to old knowledge problematized by Miller for *social* learning processes also plays a central role in the *epistemological* perspective on those processes. In section 2.1 the question if each correct mathematical knowledge is logically and consistently connected and thus deducible was discussed. How can new knowledge systematically transgress old knowledge, if it logically results from it? Here, one comes across the following paradox: On the one hand, all mathematical knowledge is logically consistent and hierarchically organized; thus new knowledge is deducible from the given foundations and does not systematically overstep the old knowledge. On the other hand, actually new and yet unknown insights in mathematics are obtained, for example, by solving problems and proving conjectures (see section 1.2).

In mathematics, a truly new construction of mathematical knowledge essentially means the production of a *new relation* between elements in the present knowledge, which is then brought into a consistent connection with the familiar knowledge. This new and more general knowledge relation significantly changes the interpretation of the old knowledge: The new knowledge is not deducible from the old knowledge, but the new knowledge gives a new interpretation to the already existing familiar knowledge and modifies it.

Referring to the four justification contexts for the argumentation and construction of mathematical knowledge in instructional communications, the problem of old and new knowledge can be characterized in the following way. The empirical and algorithmic justification contexts arguments are more deductive, thus producing no actual new mathematical knowledge. In contrast to this, situation-bound relational and independent relational justification contexts represent frames for an interactive true new construction of mathematical knowledge. Here, the „conceptual structures and relations“ are at the center of communication and of the construction of new relations. Interpreted independently, these new relations introduce new, idealized mathematical object features.

3. CLASSIFICATION OF PATTERNS OF INTERACTIVE KNOWLEDGE CONSTRUCTION

The concept of multi-layered justification contexts of children‘s mathematical argumentation and knowledge construction in the classroom, as opposed to unequivocal professional mathematical communication, together with the grid of analysis presented in section 2.3, will be used in this section for the classification of different types of interactive knowledge constructions. The four justification contexts represent the interplay between the communicable and epistemological aspects of knowledge construction. The two essential poles in the grid of analysis between which the epistemological and communicable analyses move, are:

- empirical, situated description of mathematical knowledge *and* structural, relational generality of mathematical knowledge
- mediation of the given features of the object of communication; facts, rules, logical connections *and* construction of potential interpretations of the object of communication; relations, symbols, conceptual aspects

In this section, the grid of analysis will be used so that the respective mathematical knowledge constructions are arranged in the fields of the grid on the basis of the analysis in section 3 and further qualitative justifications. This way, they precisely mark the type of knowledge construction in the span of epistemological and communicable dimension.

In the following classification, the detailed grid of analysis (cf. Fig. 17 in section 2.3) is used in a way restricted to its essential elements (cf. Fig. 112).

Communication / Epistemology	Mediation of Facts	Balance: *Facts vs. Constructions*	Construction of Interpretations
Empirical Situatedness			
Balance: *Situatedness vs. Universality*			
Relational Universality			

Figure 112. The simplified grid of analysis with its essential characteristics.

Each of the following classifications of knowledge construction is accompanied by a description of the observable *justification context*, a characterization of the respective *form of communication* and a classification of the problem of *new and old mathematical knowledge* in the present construction process (as elaborated in section 4.2).

(1) The construction and justification of mathematical knowledge by Kim and her classmate (section 3.3.1.1)

The epistemological analysis has shown that Kim (in cooperation with a classmate) develops a justification in two steps. The problem of why the magic number in a given magic square is always 66 is answered essentially in the following way. In the first justification step, Kim suggests that she does not see the magic number „66" as the result of a calculation from the „finished" number square, but she obviously uses this number for the repeated dissection into three terms of a sum in order to produce a number square with the required property. In the second argumentation step, she explains why this cannot lead to the number „77". Then she completes her explanation and names correct numbers which would lead to a square with the magic number „77". In the two steps, Kim constructs a complete, structural argument.

The analysis has shown that Kim constructs a new conceptual knowledge relation between „magic number" and „particular number square": The magic number is not only the result of a particular given square. Conversely, one can use the magic number to construct a particular square. She explains the construction proposal – which is not yet completed in all its „logical details" – for a number square with the magic number „77". The key words in her extensive argument are:

- „One divides the six-, one can divide the sixty-six into three different things. ... always. And one can do that very many times."
- „So that there are nine solutions. And if one, like, takes the twenty-two, the twenty-one, and the whatever, then the result is always sixty-six. Or if one takes the thirty-one, the sixteen and the nineteen, then the result is sixty-six again."

- „Or if one takes the thirty-one, the sixteen and the nine-, nineteen, then the result is sixty-six again."
- „... if one had seventy-seven now, one would have to put fifty-two here, for example. Then one would have to put five here, and then yet twenty."

With „divide the 66 very many times or nine times into three different things", Kim describes the construction relation for the new knowledge in a mixture of concrete numbers and general relational intentions. Then she refers to concrete numbers from the present square with the intention of arguing that „66" must then always appear as the magic number. This generalizing intention can be inferred from the remark „take ... whatever". For the – hypothetical – result „77", Kim additionally formulates – in conditional mode – how „suitable" (but arbitrary) three numbers with the sum of „77" would have to be put into a square. This is, basically, a sound, relational justification, which would have to be completed in „logical details". Thus this argument can be classified in the grid of analysis as a true construction of new knowledge (which still requires some logical details to be complete):

Communi-cation / **Episte-mology**	Mediation of Facts	Balance: *Facts vs. Constructions*	Construction of Interpretations
Empirical Situatedness			
Balance: *Situatedness vs. Universality*		●	
Relational Universality			

Figure 113. The classification of Kim's knowledge construction.

In the interaction with Kim, a situation-bound relational justification context is constituted. Kim uses not simply the concretely given properties of the number square to derive mathematical strategies, but, with the help of a new interpretation of the conceptual relation between circled numbers and magic number, she constructs new ideal features of the object „cross-out number square" in order to be able to show the required feature of the constancy of the magic number. This justification remains closely connected to the situation and is not formalized symbolically, but it is still a true mathematical argumentation. This knowledge construction is situated in a particular tension between situatedness and generality. This is suggested by means of the arrows in the diagram (Fig. 113).

From a communicable perspective, it can be stated that the signifiers communicated by Kim are open in such a way that they have to be newly interpreted within the interaction and thus that no fixed or familiar signifieds are immediately linked with these signifiers. In this way, Kim constructs true new mathematical knowledge in the interaction. This new conceptual relation is neither reduced to familiar fact or rule knowledge, nor is it separated from the familiar knowledge (for instance as isolated structures). The knowledge construction thus fulfills the criterion that it requires the old knowledge and at the same time systematically transgresses it (Miller).

(2) The construction and justification of mathematical knowledge by Judith (section 3.3.1.2)

The construction and justification of the new knowledge is again related to explaining the constancy of the magic number 66 for the given magic square. Judith developed an extensive argument, which contains the following key expressions:

- „... I could also draw another magic square onto the blackboard now. Because now, um, I'd say, these would all be very, very different numbers"
- „... then there is once the twelve, once the ten, once the eighteen, once the thirteen, once, once the nine and once the four is in the game."
- „... and none is in there twice, because I have also crossed out some of them; in one row circle only one number, otherwise, one number would be contained twice."

The epistemological and communicable analyses of this justification have shown that with her descriptions, Judith develops a generalized interpretation of the given knowledge by means of emphasizing mathematical structures and relations. Her designations mark her desire to generalize the concrete, arithmetical problem situation.

With the signifiers „another magic square", „very, very different numbers" and „numbers are in the game", Judith marks the general mathematical structure of a magic square. Then she uses the concrete numbers in the present cross-out number square. However, she uses them not to calculate, but to interpret structural relations from the point of view of her intended generalization. She constructs the relation: „Each border number is in the game once with the additive production of three circled numbers in an arbitrary cross-out number square". (The further statement, that the sum of all border numbers is also the magic number, is not explicitly thematized here). However, Judith finishes with an indirect example by using a concrete number row where one border number appears exactly once as a term of a sum with a circled number: „no border number is in there twice, because other numbers in the row are crossed out". And: „in one row circle only one number, otherwise, one border number would be contained twice". In this way, Judith develops a situation-bound, complete mathematical argument. From the given situation with concrete numbers she points at an intended generality. Using examples, she makes it clear that each of the given border numbers is in the game once, and finally she refers to the fact that no border number can appear twice in a circled number.

Judith exceeds the situation in a generalizing way, and she formulates new aspects of the mathematical object „arithmetical cross-out number square". These new constructions in the epistemological and communicable dimension do not represent arbitrary over-generalizations and remain in a perceivable reference to the problem situation with its logical structure. Judith's knowledge construction can be classified in the grid of analysis (Fig. 114) as follows:

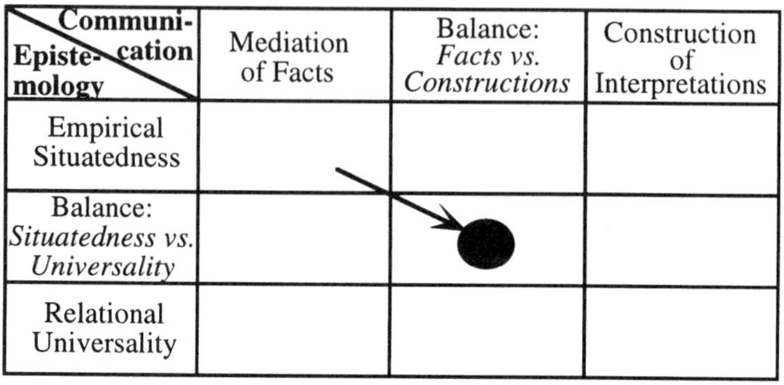

Figure 114. The classification of Judith's knowledge construction.

The interaction scene with Judith also occurs in a situation-bound relational justification context. Judith uses and extends the conceptual relation in cross-out number squares, according to which the numbers in the square result as sums of two border numbers, to the particular relation for circled numbers. Each border number is responsible exactly once for one of the circled numbers as a term of a sum. With the help of this conceptual arithmetical relation, new, idealized features of cross-out number squares are introduced and described. When doing so, Judith interprets the given features and concrete numbers as embodiments of relations, and the empirical, visual aspects are not directly taken as the key issues in the argumentation. The intended generalization in Judith's knowledge construction on the basis of the number square with the given concrete numbers is suggested in the diagram (Fig. 114) by means of the arrow.

Furthermore, one can observe that the signifiers communicated by Judith are intentionally open from a communicable point of view. This becomes obvious for example with metaphorical descriptions such as, „another magic square, all very, very different numbers" and „each border number is in the game once". These signifiers must be interpreted intentionally by the participants in the interaction, e.g. by means of seeing the embodied relations. In this way, Judith constructs true new mathematical knowledge in the interaction in the form of a new, extended conceptual relation in cross-out number squares. This relation is connected with the concrete number square and the problem, thus requires this old knowledge. On the other hand, it systematically oversteps the old mathematical knowledge by means of the intended generalizations. She uses the arithmetical relation between border numbers and numbers in the square in a new and more general sense, and she offers an indirect argumentation by means of examples that one border number cannot appear twice with circled numbers.

(3) The second construction and justification of mathematical knowledge by Kim (section 3.3.1.3)

The knowledge construction to be classified now is also related to the problem of why the magic number 66 is always the result in the present magic square. In her justification, Kim introduces a very particular description in her argument for explaining the given concrete cross-out number square with the magic number 66 as a general cross-out number square. She outlines how one can produce arbitrary cross-out number squares with the same magic number 66. Kim‘s general argument is formulated using the following key expressions:

- „... take the sixty-six apart into six numbers."
- „... the numbers, if that was cut apart now, ..."
- „... still simply lay the (numbers) there somehow ..."

Thus she constructs the general mathematical relation: „Arbitrary additive dissection of the magic number 66 into 6 terms of a sum, which serve as arbitrary border numbers". In this way, Kim formulates a construction procedure for a general cross-out number square, in which 66 (basically any given number) will result as the magic number. Thus Kim extends the structural construction possibilities for magic squares. One can consider a certain magic number chosen beforehand for constructing the border numbers. The mathematical subject „cross-out number square" is thus extended and a new relation of the mathematical knowledge is introduced. At the same time, Kim's description aims at the generality of the mathematical relations („still simply put there somehow"). The connection between the sum of the border numbers and the sum of the circled numbers is only hinted at by Kim in a general way, without making it explicit at this point.

Communication / Epistemology	Mediation of Facts	Balance: *Facts vs. Constructions*	Construction of Interpretations
Empirical Situatedness			
Balance: *Situatedness vs. Universality*		●	
Relational Universality			

Figure 115. The classification of Kim's second knowledge construction.

Thus Kim has extended the mathematical subject „cross-out number square" to a general form and developed a new interpretation and justification for the constancy of the magic number. After the teacher's inquiry, Kim subsequently concretizes the general description formulated here, without questioning her intentions of mathematical generalization. (Kim's concretization of her argument gives additional confirmations and helps in the interpretation of her first quite generally formulated justification). The new knowledge construction given by Kim can be classified in the grid of epistemological and communicable dimensions as in Figure 115.

Referring to the given situation of the number square with the particular magic number 66, Kim constructs a new conceptual relation to produce magic squares with the condition that this cross-out number square will have the magic number 66 (or intentionally an arbitrary, other magic number). Thus a situation-bound relational justification context is built in this interaction scene, and with regard to the mathematical knowledge a new, ideal object property of magic squares is described with the help of the new conceptual relation. The intended general concept relation from the given concrete circumstances is again suggested in the grid of analysis (Fig. 115) by means of the arrow.

From a communicable point of view, it can be stated that Kim communicates open signifiers yet to be interpreted in the interaction. This communicative openness becomes visible with the metaphorical descriptions: „Take the 66 apart into six numbers; then still lay the cut apart numbers somehow". No fixed signifieds can be assigned to these signifiers. They must be independently interpreted by the participants in the interaction. Thus, in this scene, Kim again constructs true new mathematical knowledge, which requires the old knowledge and the given mathematical problem and at the same time systematically oversteps it by means of the introduction of a new conceptual relation. In an open mathematical communication, Kim constructs new, situation-bound mathematical knowledge using descriptions that are adequate for children her age.

(4) The third construction and justification of mathematical knowledge by Kim (section 3.3.1.4)

The mathematical problem in this interaction scene consists in reconstructing a missing number in a cross-out number square. In the course of the lesson, Kim had already begun developing an argument in her suggestion to determine the missing number in the cross-out number square (transcript 60 – 67). First of all, the magic number is to be determined by means of three circled numbers, and then the missing number in the cross-out number square is to be calculated with the familiar magic number (cf. section 3.3.1.4). Now Kim repeats her knowledge construction and explains it. First, Kim has extensively presented how the magic number is calculated from the numbers in the diagonal (from top right to bottom left) (transcript, 106 – 145).

In Kim's proposed construction, the following key expressions can be found:

- „One <u>circles</u> the fifteen and this fifteen."
- „... adds them. And calculates <u>how much there must be up to forty-five</u>."

These key expressions contain the general argument and the new knowledge relation in a condensed form. Later in the course of interaction, this relation becomes more detailed and concretized without questioning the general character. The explanatory description expresses the new mathematical relation. Kim applies the procedure of circling three numbers and calculating the magic number in the empty field. The addition of three circled numbers is extended to a „general" addition, in which „unknown" numbers can be added as well (which is not possible in a purely arithmetical situation).

Then an important key expression appears again in Kim's concrete construction:

- „... And then one still has to calculate, how much is left up to forty-five"

Kim has constructed a new, general knowledge relation. She calculates the magic number in two ways: as the sum of three concrete, circled numbers, and, moreover, as the sum of two concrete, circled numbers and the empty field, which she has circled as well – and thus marked as a possible term of a sum for the magic number. Furthermore, this new relation contains a flexible reversibility in the mathematical structure: „With three circled numbers one can determine the magic number" is reversed to „With the magic number and two circled numbers one can determine the third circled number". Kim introduces new mathematical relations and generalizations in her knowledge construction. Additionally, she extends the ontological, mathematical aspects of the subject „cross-out number square". By means of the more exact and concrete explanation – also required by the teacher – Kim produces the „logical" and operational incorporation of her general description into the problem. Kim's knowledge construction can be represented in the grid of analysis (Fig. 116) as follows:

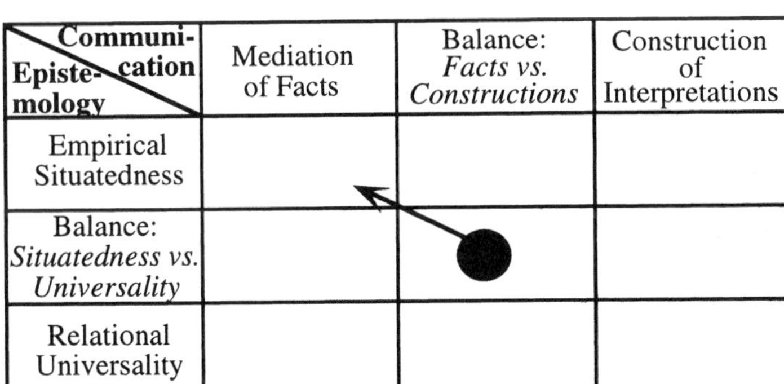

Figure 116. The classification of Kim's third knowledge construction.

The interaction scenes with Kim to determine the missing number in the magic square (see section 3.3.1.4) occur in situation-bound relational justification contexts. The epistemological analysis has shown that already in statement 60, a new conceptual knowledge relation was essentially constructed: „...one could also do it like this, that one, one would do that now already. Then and then doesn't have the number yet. And then calculates together what's missing there. And then one can also calculate how, what belongs there." The verbal representation here is very general and open and thus in need of interpretation. This interpretation is requested by the teacher, and Kim increasingly concretizes her knowledge construction. She includes more and more of the concrete, situated properties. The new knowledge relation remains open and also acts as a generally intended relation to describe new features of the object cross-out number squares: The calculation of the crossing sum from the three circled numbers is newly interpreted as an exemplary equation, and in this way, new features and dependencies for circled numbers are introduced. Kim‘s gradual concretization of her general construction proposal is represented in the diagram (Fig. 116) by an arrow.

Together with the openness of the communication, it can be observed that true new mathematical knowledge is constructed in these interaction scenes. A balance between the old knowledge and general, relational knowledge aspects becomes apparent. The arithmetical relation, which is interpreted as an exemplary equation by Kim with her descriptions, requires the old knowledge and systematically transgresses it as a new conceptual relation.

(5) The construction and justification of mathematical knowledge by Timo and Nele (section 3.3.2.1)

In this episode, children are expected to give a justification to the following mathematical problem in the learning environment of number walls: Why doesn‘t the goal number in two number walls with the goal number „360" change although different numbers appear at the border of the second row of the wall?

The epistemological and communicational analysis of the constructions and justifications given by Timo and Nele has shown that they mainly focus on the arithmetical pattern with its observable, concrete numbers (cf. section 3.3.2.1). The arithmetical relations „ten more" and „ten less" become the central points of the justification: The changes of the border numbers are subject to a „symmetry". They are not arbitrary, but the differences by „10" refer to the fact that there is a compensation. The border blocks in the second row of the upper wall can be compensated to „100", in such a way that these walls are basically „equal". The following key expressions can be identified in this episode:

- „... well, yes, ehm <u>here, here is 100</u>, 100... and <u>here is 110</u> and <u>90</u>. That <u>compensates each other</u>. There is <u>10 more</u> and there is <u>10 less</u>. Then it is <u>the same again</u>. ..."
- „Here, the tens are <u>distributed in another way</u>, ... then here <u>10 less</u> ... and there <u>10 more</u>. And here it is <u>the same then</u>."

In the argument, the students use concrete numbers and thus found the justification mainly on observable facts in the pattern of the numbers. They tend to construct the arithmetical relation „ten more" and „ten less", with which they clarify their conclusion: The respective border blocks can be compensated to „100", and thus they are not really different (and therefore the goal block ought to remain constant at „360"). An argument is constructed on the basis of arithmetical factual knowledge and simple arithmetical relations, but the soundness of this argument for the thematized problem is not sufficient (cf. the analysis in section 3.3.2.1). The interactively produced knowledge construction can be classified in the grid of analysis (Fig. 117) in the following way:

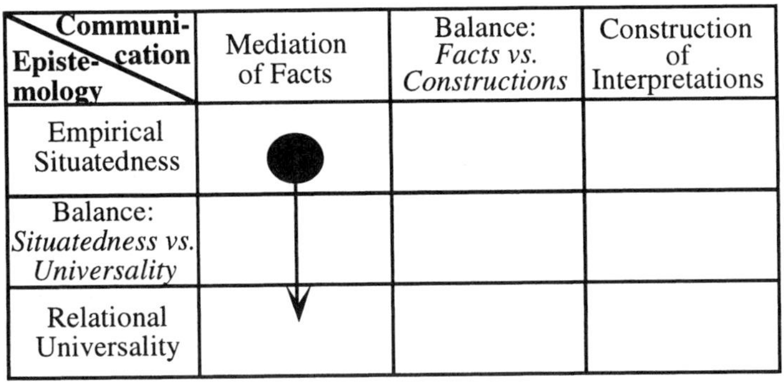

Figure 117. The classification of Timo and Nele's knowledge construction.

The interaction scene with Timo and Nele occurs in an algorithmic justification context. Children are aware of the arithmetical patterns at the surface of the number walls, and they use simple arithmetical rules and relations of „10 more" and „10 less", conceived of as arithmetical compensation. Thus no true new conceptual relation is constructed; instead, factual knowledge, which results from a direct observation of the concrete numbers in the present walls is used. The „arithmetical compensation relation" is directly connected to the observable concrete numbers. However, as a conceptual relation, which is supposed to justify the constancy of the goal block (the constancy of the number 360), it tends to separate itself from the old knowledge. This is suggested in the diagram (Fig. 117) by the arrow. The familiar knowledge which is produced in the interaction tends to disconnect itself from the old knowledge as a justification for the present problem.

From a communicational perspective it must be stated that mainly mathematical signifiers are used, that correspond to definite signifieds. The tacit statement, that the compensation of „10 more" and „10 less" between the border blocks at the same time causes the constancy of the goal block must be reconstructed from the communication, but this seems to be immediately agreed upon among the participants and does not represent true new knowledge interpretation. In the interaction, Timo and Nele do not construct new conceptual relations to extend the object features of number walls. Familiar arithmetical fact knowledge is re-produced, which however cannot justify the problem in question.

(6) The construction and justification of mathematical knowledge by Monika (section 3.3.2.2)

In this episode, the question is examined as to what are the effects of the increase of the value of the second base block (of a middle block) by „10" (marked by means of a chip) on the next levels of the number wall. The wall with the changed base block is compared with the original wall above the two number walls that is filled with all numbers. First of all, the second row of the new wall was considered. The student Monika was expected to justify the changes.

The epistemological and communicational analysis (cf. section 3.3.2.2) of Monika‘s construction and justification of knowledge has shown that after the comparison of the increases in the 5th and 6th block, she first of all formulates an explanation of why the goal block was increased by „20". In her argument, she does not use the concrete numbers. She refers to the relation „10 more" and to the fact that the number increased by „10" is calculated twice. Then she infers that also above (probably in the goal block) „that becomes 20 more". Thus she constructs an arithmetical knowledge relation, which however parallels the observable arithmetical structures and does not include underlying connections. In the second part of her argument, Monika refers to the (increased) number „90" (2nd block) and repeats the factual knowledge: The „90" is calculated twice. The essential key expressions of this interaction are the following:

- „... Because that is calculated twice ..., that is twenty more above too- ..."
- „... the ninety ... is twice. Here calculated plus fifty and then plus twenty."

The first part of the argument remains general and relational in its form. The second part of the argument becomes, on the contrary, increasingly concrete by means of taking up given numbers and related calculations. Yet no sound knowledge relation for the structure in number walls is constructed. The suggestion that the double calculation with the block raised by „10" must result in a rise by „20" „above" is not bound into the underlying logical connection of the structure of number walls. The proposed construction of a new knowledge relation is not linked to the rule structure of the arithmetical context. Thus Monika's knowledge construction can be represented in the grid of analysis (Fig. 118).

Communication / **Epistemology**	Mediation of Facts	Balance: *Facts vs. Constructions*	Construction of Interpretations
Empirical Situatedness	●		
Balance: *Situatedness vs. Universality*			
Relational Universality			●

Figure 118. The classification of Monika's knowledge construction.

In this scene, Monika argues in two different justification contexts. The first argument as to why the goal block is increased by 20, occurs in a relational justification context, which refers to several situated aspects of the present number wall. The argument uses the conceptual construction rule of number walls („The addition of two neighboring blocks results in the block above") only partially and in the form of an overgeneralization, i.e. it is immediately applied to the 4th level of the wall. Thus the conceptual relation used by Monika detaches itself from the given knowledge. In this way, Monika constructs a knowledge relation, which is no longer linked with the old knowledge, in the first part of her argument. This is indicated in the diagram by means of a black mark in the field „relational generality / construction of interpretations". The communication is open at this point, and an interpretation by the participants is necessary.

The second part of the argument tries to explain the increase of the 5th and 6th block by 10. It is situated in an empirical justification context. By means of directly pointing at and referring to the concrete numbers and by the two calculations carried out with the number 90 raised by 10, familiar factual knowledge is used. At this point, no new conceptual relation for number walls is constructed. The knowledge constructed by Monika does not transgress the familiar old knowledge. Furthermore, from a communicable point of view, the signifiers communicated in this statement are not directly linked with unequivocal signifieds. Thus Monika's argumentation expresses, on the one hand, a conceptual relation separated from the old knowledge, and repeats factual knowledge on the other. She does not construct truly new knowledge in the balance between old and new knowledge.

(7a) The construction and justification of mathematical knowledge by Jan-Pattrick (section 3.3.3.1)

In this teaching unit, children are working on arithmetical patterns in geometrical configurations for rectangular and triangular numbers. Among other things, the students are assigned the „research problem" of determining the 20th triangular number. Jan-Pattrick says that he found the 20th triangular number, namely „210". After the teacher's inquiry, he communicates his justification. First of all, Jan-Pattrick gives a description of his way of proceeding in continuation of the work on the dot patterns for the triangular numbers – obviously up to the 13th number (this is confirmed in the following episode). He applied a direct continuation of the method used so far of proceeding in the sense of a concrete production of the next triangular numbers. Triangular numbers are drawn as dot patterns and then determined by means of counting and possibly counting strategies. Then, he says, he has continued writing down addition tasks. In his explanation (transcript 84) the following key expressions can be identified:

- „Well, the thir- the fourteenth was, I believe, ninety-one plus four first calculated because it was the fourteenth. Is ninety-one plus fourteen - .... Then I have, ehm, first plus four, that were ninety-five, then plus ten, then that were one hundred and five."

The epistemological and communicational analysis (section 3.3.3.1) has shown that Jan-Pattrick very strongly pursues the concrete example calculation with the used numbers when describing his way of proceeding for the continuation of the addition tasks. He presents the calculation of the 14th triangular number, and in this example, only very vague generalizing intentions are apparent.

On first sight, the use of numbers in the calculation is dominant: „91" (here, only a very weak possible reference to the previous result of the 13th exercise becomes visible) plus „14" – here, Jan-Pattrick's hint: „Because it was the fourteenth." can be found. This could aim at the reference between the number of the exercise and the second term of the sum. Then Jan-Pattrick finishes his explanation with the description of his sum in the form of the procedure „positions extra" [i.e. adding the tens and then adding the units]. Altogether, he gives only a partial explanation of his procedure of calculating the 14th triangular number. He uses the concrete numbers mainly to carry out the calculation steps; the generalizing connections remain very vague. Thus Jan-Pattrick's description can be classified in the grid of analysis (Fig. 119) as follows:

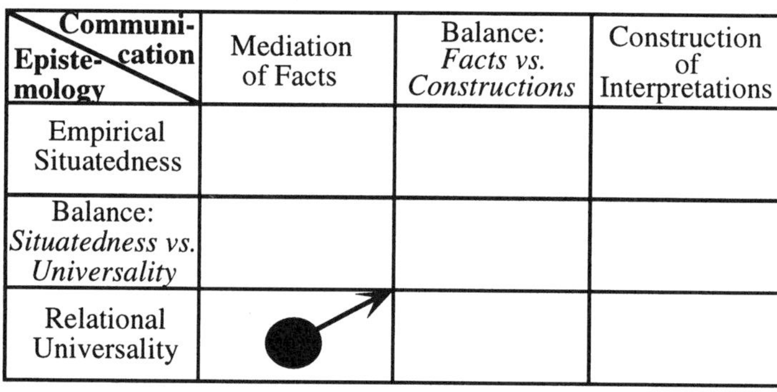

Figure 119. The classification of Jan-Pattrick's knowledge construction.

Jan-Pattrick argues in an algorithmic justification context. He partly uses familiar calculation rules, in which he includes new connections between the corresponding number of the addition task and a term of the addition task as a new conceptual relation. This new arithmetic-conceptual relation is only weakly suggested and its general character also remains unclear. Furthermore, Jan-Pattrick has now completely detached himself from the geometrical problem context and uses only the arithmetical patterns in the sequence of sums. In essence, therefore, new knowledge, increasingly detached from the old one, is produced, but a balance with the old knowledge can only be seen by way of suggestion (this is signaled by the arrow in the diagram Fig. 119). The mathematical signifiers communicated by Jan-Pattrick in the argument remain very open and must be interpreted in the interaction.

(7b) The repetition of the construction and justification of mathematical knowledge by Jan-Pattrick (section 3.3.3.1)

In this episode at the beginning of the following lesson, Jan-Pattrick's explanation of how he determined the 21st triangular number, is repeated. In order to do so, the teacher has copied Jan-Pattrick's calculation from his exercise book onto a poster in enlarged form (Fig. 81), which is now hanging on the board. The teacher asks Jan-Pattrick to explain his solution once again. In this interaction scene, Jan-Pattrick and his classmates can see the addition tasks with their results. On the one side, Jan-Pattrick names concrete calculations: „91 plus 14 (in steps) results in 105" and „190 plus the twentieth results in 210". In his repeated explanation, only several possible generalizing references to the underlying arithmetical structure become vaguely visible. The key expressions are:

- „Well, with the thirteenth, ... with the thirteenth, ninety-one was the result, with the fourteenth it was then ninety-one, I have written there plus four, that were, then I have written, equal to ninety-five, then plus ten is equal to a hundred and five. And always so on."
- „I have then calculated a hundred and ninety plus the twentieth and then I have got two hundred and ten as the result."

The epistemological analysis (section 3.3.3.1) has shown that Jan-Pattrick's explanation belongs mainly in the arithmetical-calculation context. He says how he carried out his calculations step by step. He gives only a few, implicit hints at the fact that one can always continue calculating like this. Furthermore, he only very weakly refers to the relevant connections between the previous result and the new sum (as the first term of this sum). He also names the second term of the addition task by the ordinal number of the sum.

Communication / **Epistemology**	Mediation of Facts	Balance: *Facts vs. Constructions*	Construction of Interpretations
Empirical Situatedness			
Balance: *Situatedness vs. Universality*			
Relational Universality	● ↗		

Figure 120. The classification of Jan-Pattrick's second knowledge construction.

Thus the justification aspects that are potentially developing from this representation are taken exclusively directly from the arithmetical structure. A reference to more indepth relevant justification connections is excluded. For example, the geometrical structural relations are completely ignored. Thus Jan-Pattrick's repetition of his explanation can be positioned in the grid of analysis (Fig. 120) in the same way.

The repetition of Jan-Pattrick's knowledge construction confirms his previous construction and again takes place in an algorithmic justification context. The arithmetical patterns and calculations are used and a new conceptual relation remains vague. The separation from the old knowledge, the geometric problem, is visible again. The classification of Jan-Pattrick's second knowledge construction thus corresponds to the first classification.

(8) The construction and justification of mathematical knowledge by Christopher and Nico (section 3.3.3.2)

In the previous lesson, the students examine a relation between rectangular and triangular numbers. Is a rectangular number always twice a triangular number? The corresponding poster (Fig. 85) contains the values of the first 5 rectangular numbers, their geometrical configurations and the values of the color-coded triangles within the rectangular configurations.

In the epistemological and communicatbleanalysis (section 3.3.3.2) of this justification, it became obvious that Christopher develops a continuation principle from the arithmetical pattern of the quantities noted so far, from which he determines the 6th rectangular number and the 6th triangular number. The key words in his argumentation are the following:

- „Up there it goes four. Then it goes six. Then it goes eight. And then it goes ten. [*T points in this moment at the number 20 and then to the number 30 at the left side of the table*]. Then it goes twelve [*T now points at the empty field below the number 20*]. Therefore there should be thirty-two on the other seventeen [*2 sec break*] ohm, forty-two should be on that and on the other one twenty-seven."

Christopher formulates the structural arithmetical connection using examples: 4, 6, 8, 10 and then 12. By this, he means that the difference between the visible rectangular numbers increases by „2" each time, and thus „12" must now be added to the 5th value in order to obtain the 6th rectangular number. He obviously transfers this addition of „12" to the 5th triangular number in order to determine the 6th triangular number as „27". Christopher constructs a general arithmetical relation between the rectangular numbers. He transfers this principle directly to the triangular numbers without checking the numbers already noted. The developed structural connection is inferred exclusively from the arithmetical structure, but it is not further justified, for instance by means of referring to the geometrical pattern of rectangular numbers.

With the help of a recognized general arithmetical relation, Christopher describes how one can determine the 6th rectangular number from the present numbers. This contains hidden justification aspects, which he also expresses with the description: „Therefore there should be ...". His justifying description remains actually detached from the mathematical problem of the connection between rectangular and triangular numbers – which are represented using color coding in the geometrical configuration. The arithmetical structure connection is considered in an isolated way. Thus Christopher's argument can be positioned in the grid of analysis (Fig. 121) as follows:

Communi-cation / Episte-mology	Mediation of Facts	Balance: *Facts vs. Constructions*	Construction of Interpretations
Empirical Situatedness			
Balance: *Situatedness vs. Universality*			
Relational Universality			●

Figure 121. The classification of Christopher's knowledge construction.

In this interaction scene, Christopher argues in a situation-bound relational justification context. He constructs new arithmetical relations in the given structure, which are, however, detached from the old knowledge, the geometrical problem context. The arithmetical relation between rectangular numbers is immediately transferred to the arithmetical continuation of the triangular numbers, without an additional underlying justification. The mathematical signifiers used in the communication are open and must be interactively interpreted. Thus Christopher constructs new relational knowledge, which, however, does not bear a balanced connection with the old knowledge.

Later in this episode, Christopher's justification for the 6th triangular number is corrected by Nico. Nico nominates the number „21" for this, and after the teacher's inquiry, he gives a justification (section 3.3.3.2) in which the following key expressions can be found:

- „Because twenty and twenty are forty … and one and one are two …".

The epistemological analysis (section 3.3.3.2) has shown that Nico uses the example of halving the rectangular number „42" by first halving the tens and then the unit of this number. The related intention, that the triangular number is half of the corresponding rectangular number, is not mentioned explicitly. The argument is restricted only to the procedure of the arithmetical halving. Nico states a fact of calculation.

He wants this to be understood as a justification and accordingly starts with „Because …". But the intention of understanding the calculation „20 + 20 = 40" and „1 + 1 = 2" as „the triangular number is half of the rectangular number" remains hidden. Thus Nico's justification can be classified in the grid of analysis (Fig. 122) as a communication of factual knowledge, which aims at a possible arithmetical relation. As in the previous argumentation, Nico's argument is restricted to an arithmetical pattern, without no reference to the geometrical structure.

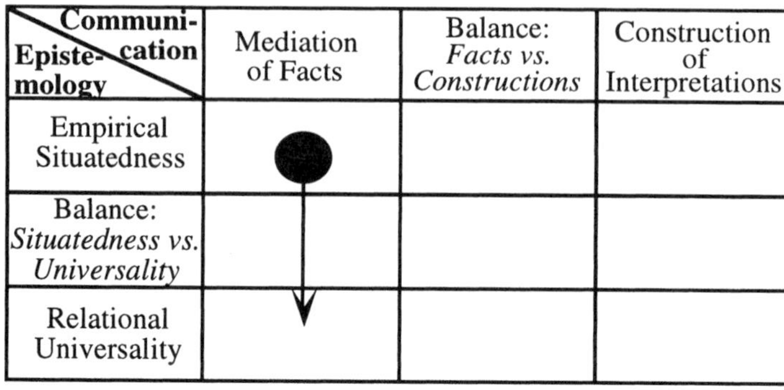

Figure 122. The classification of Nico's knowledge construction.

Nico argues within an algorithmic justification context. He communicates factual knowledge. From a communicative point of view, the communication of the calculation facts is unequivocal. The statement about the relation between rectangular and triangular numbers, which is possibly intended with it, remains open and must be interpreted. The possibility that Nico intends an arithmetical conceptual relation beyond the mere arithmetical facts has been indicated in the diagram (Fig. 122) with the arrow. Altogether, Nico does not produce true new mathematical knowledge, as his argumentation refers exclusively to arithmetical relations and does not take the geometrical knowledge problem into consideration.

(9) The construction and justification of mathematical knowledge by Rabea (section 3.3.4.1)

In this first lesson on figurate numbers, the principle of the continuation of triangular numbers is discussed. The justification, which is worked out in this episode for the acceptance of the newly constructed knowledge, is „enforced" in the form of a funnel-like interaction (cf. section 2.2) and consists of a short student contribution:

25 R Because that is the same pattern again now.

This justification of the new knowledge can only be assessed with regard to the previous interaction. As represented in detail in section 3.3.4.1, several students formulate different proposals to the problem of the continuation of the given chip configuration: an angular pattern, a double angular pattern, and a „symmetrical" triangle pattern. The students show their proposals on the black board by means of laying chips, without further additional verbal explanations. So far, only indirect conclusions can be drawn from a part of the students' proposals. These proposals can be seen as students' attempts at constructing aspects of an extended mathematical object „continuation of the chip pattern". They are situated on the empirical, concrete level, as they exhibit no traces of relational points of view. The students' first suggestions move from the given structural situation to new ontological aspects. This is seemingly justified empirically.

In the course of the interactive construction of new knowledge up to the accepted justification, the teacher comments upon and negates children's suggestions. In order to do so, she uses a sequence of similar key expressions (see section 3.3.4.1):

- „... try to lay the next one ..." (1)
- „Is that already right? ... One could think so, but it is not yet completely right ..." (5)
- „... is not yet right." (10)
- „Is that the same?" (16)
- „Oh yes, he is very close!" (22)
- „... explain why this is right now? That is right." (24)

With these descriptions, the teacher gives a hint that she has a very fixed conception of the third pattern in mind. In the interaction, the students are guided to find out which pattern is correct according to the teacher's conception. The mathematical communication is very strongly superimposed by communicative issues and then dominated by the teacher's instructional intentions (cf. section 2.2). By means of the key expressions, children are also told whether their proposals diverge from the still „secret" third configuration, pre-determined by the teacher, or correspond with it already.

Communi-cation / **Episte-mology**	Mediation of Facts	Balance: *Facts vs. Constructions*	Construction of Interpretations
Empirical Situatedness			
Balance: *Situatedness vs. Universality*			
Relational Universality			

Figure 123. The classification of the knowledge constructions related to the new triangular configuration.

For this episode with the construction of mathematical knowledge analyzed here, it can be stated that students' approaches to further development of new mathematical ontological aspects are always used by the teacher to separate them from her intended fixed geometrical pattern. Thus the interaction tends to concentrate on a finished mathematical object (kept covered) using a quasi logical incorporation into the present structure. The nature of mathematical knowledge is interpreted in an empirical way. This type of construction of mathematical knowledge can be classified in the grid of analysis (Fig. 123).

This pattern of knowledge construction emphasizes a direct, empirical interpretation and an immediate, concrete communicable description for the logical organization of knowledge. Several students offer possible solutions for the continuation of the given patterns, which contain suitable, new mathematical object aspects. (This is suggested in the diagram Fig. 123 by means of the horizontal arrow from the field „empirical situatedness / mediation of facts“ to the field „empirical situatedness / balance: facts vs. construction“.)

The teacher always confronts these suggestions with opposed, fixed knowledge features, which lead to the fact that the construction of new knowledge is reduced to the teacher's requirements (This is suggested in the diagram Fig. 123 by means of the circular arrow). The correct and accepted chip pattern is considered to be the *same* pattern, where this equality among other things, is determined by the teacher as a certain (you mean „sure“, „the right“), unequivocal triangular figure.

The argumentation between teacher and students takes place within an empirical justification context. The process of constructing the third configuration is controlled by the teacher‘s step-wise, direct „interventions”. The signifiers used in the communication are intended to be linked with unequivocal signifieds, which gradually leads to closing the communication. The fundamental fragility of the new knowledge to be constructed is removed here by means of direct intervention and therefore this new knowledge is reduced to old knowledge and to the concretely conceivable features of the first two configurations. It is a construction process, in the course of which more and more closed interactions and interpretations take place. In this process, no actual new knowledge is eventually constructed, but the conditions of the present knowledge, from which the third configuration is derived, are socially revealed in interaction

(10) The construction and justification of mathematical knowledge by Dennis (section 3.3.4.2)

This lesson looks at the continuation principle for rectangular patterns, among other things. The present interaction scene is about explaining how the third rectangular configuration can be constructed from the second one (cf. section 3.3.4.2). Following Dennis' interpretation, the teacher describes the continuation principle essentially like this: All present chips of the (second) configuration are turned over to show the red side up (this change reveals a partial pattern of the next figure) and then an angle with blue chips is added, in such a way that the next pattern appears. This general knowledge relation is what the teacher has in mind – also in Kai's explanation, she obviously assumes this way of proceeding – and she represents it again in her closing justification (transcript, 159 – 161 in section 3.3.4.2).

Kai, however, proceeds in a different way. He very strongly interprets the problem considering the empirical features of the given situation. He considers the concrete features of the third pattern and then explains how the red and the blue chips have to be added to the second configuration to construct the third one.

The description given by Kai contains concrete key expressions, which describe the situation:

- „... one <u>only</u> has to <u>do that twice</u>.”

- „... there must be here only some chips and here"
- „... and here comes another one now ..."

The teacher's subsequent explanation, in which she wants to emphasize the continuation principle, also remains restricted to the case of transition from the second to the third configuration, and in this example, possible generalizable aspects are vague and implicit. Maybe a potential generalization is hidden in the statements: „And that's why they are red now." (159) and: „Well, and what are these out here? The blue ones, what are they doing there? [*points at the angle formed by the blue chips of the third rectangular number, ...* " (159). Perhaps the intention here is to say that the current rectangle becomes the „red" part of the next rectangle and that a „blue" angle is added. In the following interaction, it remains open whether the generalizability of these statements is taken up or if the statements remain restricted to the concrete situation. In the grid of analysis (Fig. 124), the construction of new knowledge can be classified as follows:

Communication / **Epistemology**	Mediation of Facts	Balance: *Facts vs. Constructions*	Construction of Interpretations
Empirical Situatedness	●		
Balance: *Situatedness vs. Universality*			
Relational Universality			

Figure 124. The classification of Dennis' knowledge construction.

Dennis' argumentation about the construction of the third rectangular configuration from the second remains in the concrete, empirical justification context and describes a step-by-step way of proceeding, without a recognizable tendency towards generalization to subsequent configurations. Some aspects of this generalization seems to be intended by the teacher and are also possible from her explanations. Yet, in the interaction, this is not made sufficiently explicit (or taken up by the students).

Dennis does not construct new knowledge. His description refers to the visible geometrical facts. His mathematical signifiers in the form of pointing and designating are of a direct nature and refer to fixed signifieds. Thus his construction is classified in the field „empirical situatedness / mediation of facts". The teacher's knowledge construction tends to lead away from this field towards a balance between factual knowledge and potential new knowledge.

4. THREE FORMS OF MATHEMATICAL KNOWLEDGE CONSTRUCTIONS IN THE FRAME OF EPISTEMOLOGICAL AND SOCIO-INTERACTIVE CONDITIONS

Mathematical concepts are constructed as symbolic relational structures and coded in *signs and symbols*, which can be (logically, consistently) linked with each other in mathematical operations. With regard to the analysis of the conditions of construction of new mathematical knowledge in classroom interactions, mathematical signs and symbols are the central means of linking the epistemological and communicational dimension of interactive construction processes. On the one hand, signs and symbols are the *carriers of the mathematical knowledge*, and on the other hand, they are the *information in the mathematical communication* (cf. section 4.1).

According to the epistemological problem presented here, truly new mathematical knowledge can neither be seen as already familiar, logically connected coherent factual knowledge as a type of quasi-empirical knowledge nor can the new mathematical knowledge exist exclusively as a „pure relation". On the contrary, new mathematical knowledge is specifically produced and created in a context-related balance between situated, factual and rule-like circumstances of the familiar knowledge and the new knowledge structure produced by means of conceptual relations. According to the extensive analysis (section 3.3) and the classifications (section 4.3) of the documented data of teaching episodes focused on attempts of constructing and justifying new knowledge, three essential types of the intended construction of new mathematical knowledge can be distinguished. (1) Knowledge constructions, which occur almost exclusively in the frame of already familiar factual knowledge with reference to concrete, finished circumstances; (2) constructions of actual new knowledge, maintaining a relation with the contents of the given problem; (3) constructions of new general knowledge relations, detached from the contents of the given problem.

Type (1) Rule or Factual Knowledge: Knowledge construction as continuation of familiar, situated factual knowledge

With the help of the grid of analysis, one can describe this first type of the intended construction of new mathematical knowledge as follows.

On the basis of familiar, situated facts and rules, elements of new knowledge are accumulated in the form of merely rule-like or logical-deductive continuation. Possible new knowledge relations are not explicitly thematized, if they are intended at all. At most, possible new knowledge relations remain subliminally conceivable in the interactive development. This type of mathematical knowledge construction is thus pre-dominantly grounded in the factual and logical consistency of the already familiar knowledge. Thus one can say that, in the frame of such forms of knowledge construction and justification, there is no development of actual new knowledge, which would fundamentally exceed the already familiar knowledge. This form of interactive knowledge construction occurred in the episodes (5) „Timo and Nele", (6) „Monika" (second argument), (8) „Nico", (9) „Rabea and teacher" and (10) „Dennis".

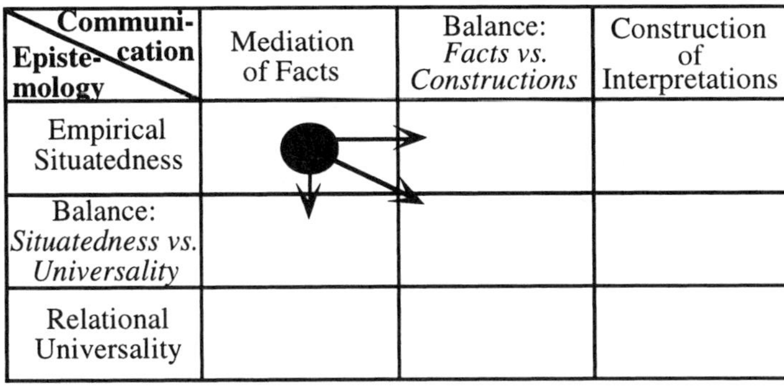

Figure 125. The classification of the construction of rule or factual knowledge.

Type (2) Problem-Related Relational Knowledge: *Knowledge relations as a balance between consistent base knowledge and new knowledge relations*

This second type of intended mathematical knowledge constructions is a „bridge" or balance between the new, relational knowledge construction and the already familiar base knowledge, which is always specific and interactively produced in the frame of context-determined formulations and features. This type obtains the following general representation in the grid of analysis. In order to identify this type of knowledge construction, it is important that by means of the newly constructed knowledge relation, the already familiar, old knowledge can also be interpreted in a new way. Thus the new knowledge is first constructed relatively independently from the familiar knowledge and systematically oversteps it, but is then also put in a connection with it in an ultimately consistent way, i.e. it also requires the old knowledge (Miller 1986).

In the frame of the theoretical analysis dimensions used here, this second type of intended mathematical knowledge relations represents successful interactive constructions and justifications of new mathematical knowledge, understood not as logically formal or abstract constructions, but occurring in elementary school instruction in appropriate specific formulations and situations. This form of interactive knowledge construction could be identified for example in the episodes (1) „Kim and classmate", (2) „Judith", (3) „Kim's second knowledge construction" and (4) „Kim's third knowledge construction" (Fig. 126).

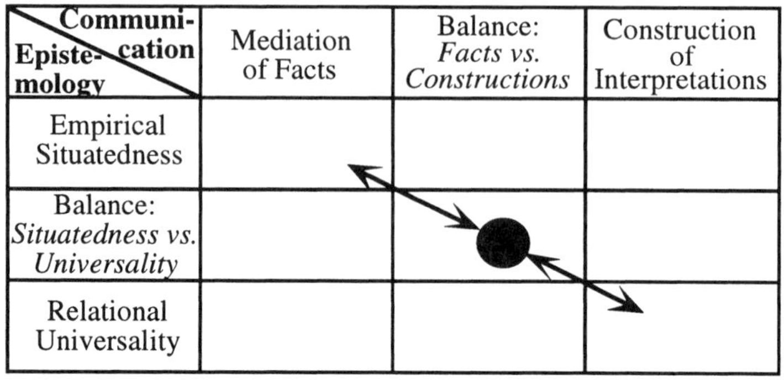

Figure 126. The classification of the construction of problem-related relational knowledge.

Type (3) Problem-Separated Relational Knowledge: *Knowledge construction as introduction of isolated knowledge structures*

The third type of intended mathematical knowledge relations is characterized by a tendency to construct general structural connections as the basis for the new knowledge. But these relational structures then lack the references to the context of the present base and factual knowledge. The „new" knowledge eventually no longer needs the old knowledge (Miller, 1986). In the grid of analysis, this construction type can be characterized as follows (Fig. 127).

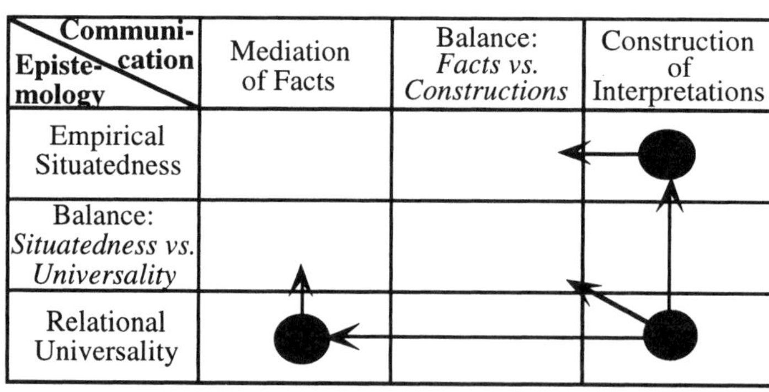

Figure 127. The classification of the construction of problem-separated relational knowledge.

The newly constructed relations remain isolated and without connection to the present knowledge or one can often state external „overgeneralizations" of structures, which allow making superficial justifications, but which are not connected to the original problem in a deeper way. This type of mathematical knowledge construction may contain an isolated mathematical structural, operational relation as well as an isolated relational interpretation of the new knowledge object. A new general mathematical relation is constructed, without being connected to the old knowledge to be able to see the old knowledge in a new light. Instances of the construction of problem-separated mathematical knowledge could be found in the episodes (6) „Monika" (first argument), (7) „Jan-Pattrick" and (8) „Christopher".

These three forms of mathematical knowledge construction in the frame of epistemological and socio-interactive conditions represent a concentrate focus on essential characteristics. This implies a pointed and general view of the knowledge constructions in the interaction scenes. This may not be considered in an isolated manner or as the only result of the investigation. Such a focused perspective always requires careful epistemology-oriented analyses as essential for completion, leading to the extensive case studies for the single interaction scenes, and thus trying to partly do justice to the observable complexity.

The extensive epistemology-oriented analyses together with the classification of the interactive knowledge constructions describe the conditions in which students' mathematical communication and interaction can take place. There exists a variety of communication frames, contrary to the unequivocal professional argumentation of research mathematicians. This communication variety contains epistemological aspects of the (school-) mathematical knowledge treated in the interaction as well as the communication problems linked with it, as they come into play especially in the context of teaching and learning.

The possibility and necessity of constructing truly new mathematical knowledge already in the early stages of mathematics teaching presents a big challenge. Usually, the learning of factual knowledge and of fixed arithmetical rules and procedures dominates elementary school mathematics instruction. If the children are given suitable substantial mathematical learning environments (see section 3.1) to work on, as in the frame of the research presented, the spectrum of their forms of justifying and of ways of constructing mathematical knowledge is extended in a significant way. Under such circumstances, also interaction scenes taken from elementary school instruction represent fruitful cases for scientific examination. Especially the mathematical communication among early learners contains a greater variability than the mathematical communication in high school years, where forms of communication are increasingly standardized and become dominated by „formalized" ways of treating (school-) mathematical knowledge. The construction and justification of truly new mathematical knowledge must occur in the situated contexts of learning environments which are accessible to children, and it must take into consideration the particular forms of communication in elementary instruction. These epistemological and communicative border conditions and relations between them represent the core of the challenge. They refer to the complexity of the problem of instructional construction of new mathematical knowledge. The epistemology-oriented analyses of interactive knowledge constructions represent an instrument for both investigating this complexity and becoming more sensitive to it in observing and assessing processes of mathematics teaching and learning.

LOOKING BACK

The general initial question of this book is: How can everyday mathematics teaching be – as properly as possible – described, understood and developed as a teaching and learning environment in which the students gain mathematical insights and increasing mathematical competence by means of the teacher's initiatives, offers and challenges? How can the „quality" of mathematics teaching be realized and appropriately described? And the following more specific research question was investigated in this book: How is new mathematical knowledge interactively constructed in a typical instructional communication among students together with the teacher? In order to answer this question at least in certain important aspects, an attempt has been made to enter as indepth as possible under the surface of the visible phenomena of the observable everyday teaching events. In order to do so, theoretical views about mathematical knowledge and communication were elaborated, which then served as the foundation for the epistemologically oriented, qualitative analysis.

At the center of this research are everyday mathematical interaction and communication processes. The following problem emerges: What do „true" mathematical instruction communications consist of, how can they be characterized and found within instruction processes? Two essential dimensions of the analysis of mathematical interactions have been used and put into a relation to each other: The epistemology of mathematical knowledge, as an historically grown theoretical knowledge domain, as the product of mathematical research and as the object of mathematical teaching and learning processes. Furthermore, the conditions of general communication in the relation to classroom communication have been described and included in the qualitative analyses.

In the practice of researching mathematicians, a true mathematical communication is uniform and aims at a consensus, which is ultimately forced by the epistemological conditions of mathematics. In the construction and justification of new mathematical knowledge, mathematicians assume elementary conceptual structures (defining equations, axioms, etc.), with whose help the ideal mathematical object is indirectly defined and then developed during the unfolding of the mathematical theory. In contrast to this, true mathematical communication during instruction is subject to various difficult conditions. First of all, it is not uniform, and often, a consensus is mainly reached under the conditions of the dominating instructional communication. In most cases, a true mathematical communication cannot be presented already at the start of instruction, as it must be slowly and jointly developed in the context of the culture of mathematics teaching.

The characterization of the conditions for a true mathematical communication within the culture of everyday mathematics teaching is an important outcome of this study. The conditions, which the mathematical communication with students during instruction is subject to, are contrasted in the poles of the two dimensions of the analysis grid. Mathematical constructions and knowledge justifications often swing back and forth between the poles of these two dimensions. On the one hand, knowledge constructions often remain on the level of mere factual knowledge and an interpretation of the knowledge, which is closely linked to the concrete case. On the other hand, constructions of arbitrary structures and relations, which are not further justified, can often be observed.

True mathematical communications with appropriate knowledge constructions require the maintenance of a balance between situatedness and intended generality in the instruction, as well as the communication of given facts and properties with the intention of an interpretation and construction of relations. The central difficulty for the success of mathematical communications during teaching and learning consists in the fact that the students' mathematical communication ability must still be slowly developed and that classroom communication may not simply be equated to the uniform professional mathematical communication. This results in the extremely demanding challenge of always maintaining the difficult balance regarding the epistemology as well as the communication of the knowledge in mathematics teaching.

The goal of this study was to describe the difficult prerequisites and conditions for successful, true mathematical communications within everyday instruction as precisely as possible and to present justificatory examples of successful mathematical classroom communications with their particularities in the analysis of selected teaching episodes. The examples of rather less successful communications additionally illustrate the conditions for true instructional communication processes.

It is a common educational tradition, that when observing rather unsuccessful and also very successful mathematical learning or teaching processes, effective measures of change and improvement of the discovered inadequacies are demanded and often quickly implemented. In most cases, this wish is based on the assumption that necessary improvements of learning, understanding and instructing processes are immediately and directly possible. In contrast to this, the present study has shown that a mathematical communication process must be seen as an autonomous culture which to a certain extent, is subject to its own inherent laws and general parameters, and which one therefore cannot directly influence from the outside in order to initiate supposed changes.

A careful, epistemologically oriented analysis of mathematical interaction processes, which allows for going deeper into the complex events of interactive mathematical knowledge constructions with the help of fundamental theoretical perspectives and developed analysis instruments, is necessary in order to discover rules and connections within the autonomous mathematical instruction culture. Without the knowledge of these hidden connections the realization of changes in mathematical interaction processes remains an experiment according to the motto, „trial and error". Direct, massive interventions from the outside into the relatively autonomous mathematical interaction process by the didactical researcher are not possible and ultimately, they cannot have an effect for positive change. Changes with accompanying improvements to a true mathematical teaching culture can ultimately only start from the inside and spread slowly.

Realistic initiations of changing processes require the common work and reflection of teachers and scientists, thus a co-operation between educational theory and instructional practice. For this as well, the analyses of mathematical teaching communications such as carried out in this book are an indispensable basis, because essential characteristics for successful mathematical communications under the conditions of learning and everyday mathematics instruction are elaborated. But even on this basis intended changes cannot be „dictated".

Intended changes especially concern the autonomous (autopoietic) communication system. The complex mathematical interaction, however, cannot be changed simply by „rules of conduct“ or „instructions for use“ and adjusted to „good“ behavior. In most cases, interactions can only be influenced indirectly and stimulated to produce slowly developing alternatives.

Such positive changes must first of all start in the teachers' personal attitudes. The personal attitude may not be seen as a permanently explicit state of consciousness of the teacher. Instead, a personal, professional attitude is a behavioral pattern which has passed into the subconscious and which also becomes effective in the course of mathematical interactions, in which the teacher is completely involved, and influences and perhaps regulates the communication.

At this point, possible processes of change and lately also of improvement to a true mathematical teaching communication can begin. Their realization requires a careful analysis and elaboration of theoretical concepts about the teacher's forms of communication and intervention together with the students' interactive reactions, one can observe in everyday mathematics teaching. Such observations ought to be collected and compared in successful as well as in less successful mathematical instruction communications. Subsequently, a research design for an empirical examination ought to be planned and after the ending of the empirical experiment, a careful analysis ought to be carried out. But this would be a new and extensive research project in mathematics education, which would need to be presented in another study.

REFERENCES

Baraldi, C., Corsi, G. & Esposito, E. (1997). *GLU. Glossar zu Niklas Luhmanns Theorie sozialer Systeme*. Frankfurt am Main: Suhrkamp.

Bauersfeld, H. (1978). Kommunikationsmuster im Mathematikunterricht – Eine Analyse am Beispiel der Handlungsverengung durch Antworterwartung. In H. Bauersfeld (Ed.), *Fallstudien und Analysen zum Mathematikunterricht* (pp. 158 – 170). Hannover: Schroedel.

Bauersfeld, H. (1982). Analysen zur Kommunikation im Mathematikunterricht. In H. Bauersfeld , H. W. Heymann, G. Krummheuer, J. H. Lorenz & V. Reiß (Eds.), *Analysen zum Unterrichtshandeln* (pp. 1-40). Köln: Aulis.

Bauersfeld, H. (1983). Subjektive Erfahrungsbereiche als Grundlage einer Interaktionstheorie des Mathematiklernens und -lehrens. In H. Bauersfeld, H. Bussmann, G. Krummheuer, J. H. Lorenz & J. Voigt (Eds.), *Lernen und Lehren von Mathematik* (pp. 1 - 56). Köln: Aulis.

Bauersfeld, H. (1988). Interaction, Construction and Knowledge: Alternative Perspectives for Mathematics Education. In D. A. Grouws, T. J. Cooney & D. Jones (Eds.), *Effective Mathematics Teaching* (pp. 27 – 46). Reston, Virginia: NCTM & Lawrence Erlbaum.

Bauersfeld, H. (1994). Theoretical Perspectives on Interaction in the Mathematics Classroom. In R. Biehler, R. W. Scholz, R. Sträßer & B. Winkelmann (Eds.), *Didactics of Mathematics as a Scientific Discipline* (pp. 133 – 146). Dordrecht: Kluwer Academic Publishers.

Bauersfeld, H. (1995). „Language Games" in the Mathematics Classroom: Their Function and Their Effects. In P. Cobb & H. Bauersfeld (Eds.), *The Emergence of Mathematical Meaning – Interaction in Classroom Cultures* (Vol. 2, pp. 271 – 291). Hillsdale, New Jersey: Lawrence Erlbaum.

Bauersfeld, H., Krummheuer, G. & Voigt, J. (1988). Interactional Theory of Learning and Teaching Mathematics and related Microethnographical Studies. In H. G. Steiner & A. Vermandel (Eds.), *Foundations and Methodology of the Discipline Mathematics Education* (pp. 174 – 188). Antwerpen.

Bekemeier, B. (1987). *Martin Ohm (1792 – 1872). Universitäts- und Schulmathematik in der neuhumanistischen Bildungsreform*. Göttingen: Vandenhoeck & Ruprecht.

Bereiter, C. (1985). Towards a Solution of the Learning Paradox. *Review of Educational Research, 55*(2), 201 – 226.

Bereiter, C. (1994). Constructivism, Socioculturalism, and Popper's World 3. *Educational Researcher, 23*(7), 21 – 23.

Bereiter, C. (1997). Situated Cognition and How to Overcome It. In D. Krishner & J. A. Whiston (Eds.), *Situated Cognition. Social, Semiotic, and Psychological Perspectives.* (pp. 281 – 300). Mawah, New Jersey: Lawrence Erlbaum Associates, Publishers.

Bickhard, M. H. (1991). The Import of Fodor's Anti-Constructivist Argument. In L. P. Steffe (Ed.), *Epistemological Foundations of Mathematical Experiences* (pp. 14 – 25). New York: Springer.

Bishop, A. J. (1988). *Mathematical Enculturation. A Cultural Perspective on Mathematics Education.* Dordrecht: Kluwer Academic Publishers.

Borel, E. (1965). *Elements of the theory of probability*. New Jersey: Englewood.

Bourbaki, N. (1971). The Architecture of Mathematics. In F. Le Lionnais (Ed.), *Great Currents of Mathematical Thought* (Vol. I, pp. 23 – 36). New York: Dover Publications, Inc.

Bromme, R., Seeger, F. & Steinbring, H. (1990). Aufgaben, Fehler und Aufgabensysteme. In R. Bromme, H. F. Seeger & Steinbring (Eds.), *Aufgaben als Anforderungen an Lehrer und Schüler* (pp. 1 – 30). Köln: Aulis.

Bromme, R. & Steinbring, H. (1990a). Die epistemologische Struktur mathematischen Wissens im Unterrichtsprozeß. Eine empirische Analyse von vier Unterrichtsstunden in der Sekundarstufe I. In R. Bromme, F. Seeger & H. Steinbring (Eds.), *Aufgaben als Anforderungen an Lehrer und Schüler* (pp. 151 – 229). Köln: Aulis.

Bromme, R. & Steinbring, H. (1990b). Eine graphische Analysetechnik für Unterrichtsverläufe. In K. Haussmann & M. Reiss (Eds.), *Mathematische Lehr-Lern-Denkprozesse*. (Vol. 9, pp. 55 – 81). Göttingen: Hogrefe.

Bromme, R. & Steinbring, H. (1994). Interactive Development of Subject Matter in the Mathematics Classroom. *Educational Studies in Mathematics, 27*(3), 217 – 248.

Brosseau, G. (1997). *Theory of Didactical Situations in Mathematics*. Dordrecht: Kluwer Academic Publishers.

Brosseau, G. & Otte, M. (1991). The Fragility of Knowledge. In A. Bishop, S. Mellin-Olsen & J. v. Dormolen (Eds.), *Mathematical Knowledge: Its Growth Through Teaching* (pp. 13 – 36). Dordrecht: Kluwer Academic Publishers.

Bruner, J. (1972). Einige Elemente des Entdeckens. In A. Halbfas et al., (Ed.), *Entwicklung der Lernfähigkeit.* (pp. 84 – 99). Stuttgart: Klett.

Cassirer, E. (1923). *Substance and Function & Einstein's Theory of Relativity.* New York, Dover.

Cassirer, E. (1957). *The Philosophy of Symbolic Forms. Volume 2: Mythical Thought.* New Haven, Conn., Yale University Press.

Cassirer, E. (1957). *The Philosophy of Symbolic Forms. Volume 3: The Phenomenology of Knowledge.* New Haven, Conn., Yale University Press.

Changeux, J.-P. & Connes, A. (1999). *Conversations on Mind, Matter, and Mathematics.* Princeton, New Jersey: Princeton University Press.

Cobb, P. (1986). Concrete can be Abstract: A Case Study. *Educational Studies in Mathematics, 17*, 37 – 48.

Cobb, P. (1994). Where is the Mind? Constructivist and Sociocultural Perspectives on Mathematical Development. *Educational Researcher, 23*(7), 13 – 20.

Cobb, P. & Bauersfeld, H. (Eds.). (1995). *The Emergence of Mathematical Meaning – Interaction in Classroom Cultures* (Vol. 2). Hillsdale, New Jersey: Lawrence Erlbaum.

Cobb, P. & Yackel, E. (1998). A Constructivist Perspective on the Culture of the Mathematics Classroom. In F. Seeger, J. Voigt & U. Waschescio (Eds.), *The Culture of the Mathematics Classroom* (pp. 158 – 190). Cambridge, UK: Cambridge University Press.

Cobb, P., Yackel, E. & Wood, T. (1991). Learning through Problem Solving: A Constructivist Approach to Second Grade Mathematics. In E. v. Glasersfeld (Ed.), *Radical Constructivism in Mathematics Education* (pp. 157 – 176). Dordrecht, Boston, London: Kluwer Academic Publishers.

Davis, P. J. & Hersh, R. (1981). *The Mathematical Experience.* Boston: Birkhäuser.

Dehaene, S. (1997). *The Number Sense. How the Mind Creates Mathematics.* London: Allen Lane The Penguin Press.

Dörfler, W. (1995). Mathematical Objects, Representations, and Imagery. In R. Sutherland & J. Mason (Eds.), *Exploiting Mental Imagery with Computers in Mathematics Education* (pp. 82 – 94). Berlin: Springer.

Duval, R. (2000). Basic Issues for Research in Mathematics Education. In T. Nakahara & M. Koyama (Eds.), *Proceedings of the 24thInternational Conference for the Psychology of Mathematics Education* (Vol. I, pp. 55 – 69). Hiroshima, Japan: Nishiki Print Co., Ltd.

Eccles, P. J. (1997). *An Introduction to Mathematical Reasoning. Numbers, Sets, and Functions.* Cambridge: Cambridge University Press.

Frege, G. (1969). *Funktion, Begriff, Bedeutung. Fünf logische Studien.* Göttingen: Vandenhoeck & Ruprecht.

Freudenthal, H. (1973). *Mathematics as an Educational Task.* Dordrecht: D. Reidel Publishung Company.

Glasersfeld, E. v. (Ed.). (1991). *Radical Constructivism in Mathematics Education.* Dordrecht, Boston, London: Kluwer Academic Publishers.

Goodman, N. (1979). Mathematics as an Objective Science. *American Mathematical Monthly, 86*, 540 – 551.

Hacking, I. (1975). *The emergence of probability.* Cambridge: Cambridge University Press.

Harten, G. v. & Steinbring, H. (1985). *Aufgabensysteme im Stochastikunterricht – Didaktische Konzepte zur Materialentwicklung und Lehrerfortbildung im LEDIS-Projekt.* Bielefeld: Occasional Paper Nr. 71 des IDM.

Heintz, B. (2000). *Die Innenwelt der Mathematik. Zur Kultur und Praxis einer beweisenden Disziplin.* Wien New York: Springer.

Hersh, R. (1997). *What is Mathematics, Really?* London: Jonathan Cape.

Hoffmann, M. (2000). Die Paradoxie des Lernens und ein semiotischer Ansatz zu ihrer Auflösung. *Zeitschrift für Semiotik, 22*(1), 31 - 50.

Indurkhya, B. (1994). Metaphor as Change of Representation: An Interaction Theory of Cognition and Metaphor. In J. Hintikka (Ed.), *Aspects of Metaphor* (pp. 95 – 150). Dordrecht: Kluwers Academic Publishers.

Jahnke, H. N. (1978). *Zum Verhältnis von Wissensentwicklung und Begründung in der Mathematik – Beweisen als didaktisches Problem. Materialien und Studien Bd. 10.* Bielefeld: IDM Universität Bielefeld.

Jahnke, H. N. (1990). J.F.Herbart (1776 – 1841): Nach-Kantische Philosophie und Theoretisierung der Mathematik. In G. König (Ed.), *Konzepte des mathematisch Unendlichen im 19. Jahrhundert* (pp. 165 – 188). Göttingen: Vandenhoeck & Ruprecht.

Jahnke, H. N. & Otte, M. (1981). On 'Science as a Language'. In H. N. Jahnke & M. Otte (Eds.), *Epistemological and Social Problems of the Sciences in the Early Nineteenth Century*. Dordrecht: Reidel.

Jungwirth, H. (1994). Aufgabenlöseprozesse in Kleingruppen von Schülerinnen und Schülern. In H. Maier & J. Voigt (Eds.), *Verstehen und Verständigung im Mathematikunterricht – Arbeiten zur interpretativen Unterrichtsforschung* (pp. 112 – 143). Köln: Aulis.

Jungwirth, H., Steinbring, H., Voigt, J. & Wollring, B. (1994). Interpretative Unterrichtsforschung in der Lehrerbildung. In H. Maier & J. Voigt (Eds.), *Verstehen und Verständigung im Mathematikunterricht – Arbeiten zur interpretativen Unterrichtsforschung* (pp. 12 – 42). Köln: Aulis.

Keyser, C. J. (1932/33). Mathematics as a Culture Clue. *Scripta Mathematica, 1*, 185 – 203.

KM, K. d. L. N.-W. (1985). *Richtlinien und Lehrpläne für die Grundschule in Nordrhein-Westfalen: Mathematik*. Düsseldorf.

Knorr-Cetina, K. D. (1981). *The Manufacture of Knowledge. An Essay on the Constructivist and Contextual Nature of Science*. Oxford: Pergamon Press.

Krainer, K. (1990). *Lebendige Geometrie, Überlegungen zu einem integrativen Verständnis von Geometrieunterricht anhand des Winkelbegriffs*. Frankfurt a. M.: Peter Lang.

Krauthausen, G. (1995). Zahlenmauern im zweiten Schuljahr. Ein substantielles Aufgabenformat. *Grundschulunterricht, 42*, 5 – 9.

Krauthausen, G. (1996). Lehren, Lernen & Lehren lernen – ein Beitrag zur 2. Phase der Lehrerbildung. In K. P. Müller (Ed.), *Beiträge zum Mathematikunterricht 1996* (pp. 253 – 256). Hildesheim: Franzbecker.

Krummheuer, G. (1984). Zur unterrichtsmethodischen Diskussion von Rahmungsprozessen. *Journal für Mathematik Didaktik, 5*(4), 285 – 306.

Krummheuer, G. (1988). Verständigungsprobleme im Mathematikunterricht. *Der Mathematikunterricht, 34*(2), 55 – 60.

Krummheuer, G. (1997). Zum Begriff der »Argumentation« im Rahmen einer Interaktionstheorie des Lernens und Lehrens von Mathematik. *Zentralblatt für Didaktik der Mathematik*(1), 1 – 10.

Krummheuer, G. (1998). Formats of Argumentation in the Mathematics Classroom. In H. Steinbring, M. G. Bartolini Bussi & A. Sierpinska (Eds.), *Language and Communication in the Mathematics Classroom* (pp. 223 – 234). Reston, Virginia: National Council of Teachers of Mathematics.

Krummheuer, G. & Voigt, J. (1991). Interaktionsanalysen von Mathematikunterricht – Ein Überblick über Bielefelder Arbeiten. In H. Maier & J. Voigt (Eds.),*Interpretative Unterrichtsforschung* (pp. 13 – 32). Köln: Aulis.

Lakoff, G. & Núñez, R. E. (2000). *Where Mathematics Comes from. How the Embodied Mind Brings Mathematics into Being*. New York: Basic Books.

Lave, J. (1988). *Cognition in Practice. Mind, Mathematics and Culture in Everyday Life*. Cambridge: Cambridge University Press.

Lave, J. & Wenger, E. (1991). *Situated Learning – Legitimate Peripheral Participation*. Cambridge: Cambridge University Press.

Lawler, R. W. (1990). Constructing Knowledge from Interactions. *The Journal of Mathematical Behavior, 9*(2), 177 – 191.

Loève, M. (1978). Calcul des probabilités. In J. Dieudonné (Ed.), *Abgrégé d'histoire des mathématiques 1700 – 1900* (Vol. II, pp. 277 – 313). Paris: Hermann.

Lorenz, J. H. (1995). Arithmetischen Strukturen auf der Spur. Die Funktion und Wirkungsweise von Veranschaulichungsmitteln. *Die Grundschulzeitschrift, Heft 28* (März), 9 – 12.

Luhmann, N. (1982). *Liebe als Passion. Zur Codierung von Intimität*. Frankfurt am Main: Suhrkamp.

Luhmann, N. (1996). Takt und Zensur im Erziehungssystem. In N. Luhmann & K.-E. Schorr (Eds.), *Zwischen System und Umwelt. Fragen an die Pädagogik.* (pp. 279 – 294). Frankfurt am Main: Suhrkamp.

Luhmann, N. (1997). *Die Gesellschaft der Gesellschaft*. Frankfurt am Main: Suhrkamp.

Luhmann, N. (1997). Was ist Kommunikation? In F. B. Simon (Ed.), *Lebende Systeme. Wirklichkeitskonstruktionen in der systemischen Therapie.* (pp. 19 – 31). Frankfurt am Main: Suhrkamp.

Maier, H. & Steinbring, H. (1998). Begriffsbildung im alltäglichen Mathematikunterricht – Darstellung und Vergleich zweier Theorieansätze zur Analyse von Verstehensprozessen *Journal für Mathematik-Didaktik*, 292 – 329.

Maier, H. & Voigt, J. (Eds.). (1991). *Interpretative Unterrichtsforschung*. Köln: Aulis.
Maistrov, L. E. (1974). *Probability Theory: A Historical Sketch*. New York: Academic press.
Mason, J. & Pimm, D. (1984). Generic Examples: Seeing the General in the Particular. *Educational Studies in Mathematics, 15*, 277 – 289.
Maturana, H. R. & Varela, F. J. (1987). *Der Baum der Erkenntnis. Die biologischen Grundlagen des menschlichen Erkennens*. Bern, München: Scherz.
Mehan, H. (1979). *Learning Lessons*. Cambridge, Mass.: Harvard University Press.
Miller, M. (1986). *Kollektive Lernprozesse. Studien zur Grundlegung einer soziologischen Lerntheorie*. Frankfurt: Suhrkamp Verlag.
Mises, von R. (1972). Wahrscheinlichkeit, Statistik und Wahrheit. Wien: Springer.
Neth, A. & Voigt, J. (1991). Lebensweltliche Inszenierungen – Die Aushandlung schulmathematischer Bedeutungen an Sachaufgaben. In H. Maier & J. Voigt (Eds.), *Interpretative Unterrichtsforschung* (pp. 79 – 116). Köln: Aulis.
Nickson, M. (1994). The Culture of the Mathematics Classroom: An Unknown Quantity? In S. Lerman (Ed.), *Cultural Perspectives on the Mathematics Classroom* (pp. 7 – 35). Dordrecht: Kluwer Academic Publishers.
Nöth, W. (2000). *Handbuch der Semiotik*. Stuttgart Weimar: Verlag J.B. Metzler.
Núñez, R. E. (2000). Mathematical Idea Analysis: What Embodied Cognitive Science Can Say about the Human Nature of Mathematics. In T. Nakahara & M. Koyama (Eds.), *Proceedings of the 24th International Conference for the Psychology of Mathematics Education* (Vol. I, pp. 3 – 22). Hiroshima, Japan: Nishiki Print Co., Ltd.
Ogden, C. K. & Richards, F. A. (1923). *The Meaning of Meaning*. London: Routledge & Kegan.
Otte, M. (1984). *Was ist Mathematik?* Bielefeld: Occasional Paper Nr. 43 des IDM.
Otte, M. (2001). *Mathematical Epistemology from a Semiotic Point of View*. Paper presented to the Discussion Group on Semiotics in Mathematics Education Research at the 25th PME International Conference, The Netherlands, University of Utrecht, July 12 – 17, 2001. Unpublished manuscript.
Presmeg, N. (2001). *Progressive Mathematizing Using Semiotic Chaining*. Paper presented to the Discussion Group on Semiotics in Mathematics Education Research at the 25th PME International Conference, The Netherlands, University of Utrecht, July 12 – 17, 2001. Unpublished manuscript.
Radford, L. (2001). *On the Relevance of Semiotics in Mathematics Education*. Paper presented to the Discussion Group on Semiotics in Mathematics Education Research at the 25th PME International Conference, The Netherlands, University of Utrecht, July 12 – 17, 2001. Unpublished manuscript.
Radford, L. (2003). On Culture and Mind: A Post-Vygotskian Semiotic Perspective. With an Example From Greek Mathematical Thought, in: M. Anderson, A. Sáenz – Ludlow, S. Zellweger, and V. Cifarelli (eds.), *Educational Perspectives on Mathematics as Semiosis: From Thinking to Interpreting to Knowing*, Legas Publishing, Ottawa, pp. 49 – 79.
Resnik, M. D. (2000). Some Remarks on Mathematical Progress from a Structural Perspective. In E. Grossholz & H. Berger (Eds.), *The Growth of Mathematical Knowledge* (pp. 353 – 362). Dordrecht: Kluwer Academic Publishers.
Rotman, B. (1987). *Signifying Nothing – The Semiotics of Zero*. Stanford, California: Stanford University Press.
Rotman, B. (1988). Toward a Semiotics of Mathematics. *Semiotica, 72*(1/2), 1 – 35.
Sáenz – Ludlow, A. (2001). *Classroom Mathematics Discourse as an Interpreting Game*. Paper presented to the Discussion Group on Semiotics in Mathematics Education Research at the 25th PME International Conference, The Netherlands, University of Utrecht, July 12 – 17, 2001. Unpublished manuscript.
Saussure, F. d. (1967). *Grundfragen der allgemeinen Sprachwissenschaft*. Berlin: Walter de Gruyter & Co.
Scherer, P. (1997). Substantielle Aufgabenformate – jahrgangsübergreifende Beispiele für den Mathematikunterricht, Teile I bis III. *Grundschulunterricht* (1, 4 and 6), 34 – 38, 36 – 38 and 54 – 56.
Searle, J. R. (1995). *The Construction of Social Reality*. New York: The Free Press.
Sinclair, J. & Coulthard, R. (1975). *Towards an Analysis of Discourse: The English used by Teachers and Pupils*. London: Oxford University Press.
Solomon, Y. (1989). *The Practice of Mathematics*. London: Routledge & Kegan.
Steinbring, H. (1980). *Zur Entwicklung des Wahrscheinlichkeitsbegriffs – Das Anwendungsproblem der Wahrscheinlichkeitstheorie aus didaktischer Sicht*. Bielefeld: Materialien und Studien Bd. 18 des IDM Bielefeld.

Steinbring, H. (1989). Routine and Meaning in the Mathematics Classroom. *For the Learning of Mathematics, 9*(1), 24 – 33.

Steinbring, H. (1991a). Eine andere Epistemologie der Schulmathematik – Kann der Lehrer von seinen Schülern lernen? *mathematica didactica, 14*(2/3), 69 – 99.

Steinbring, H. (1991b). Mathematics in Teaching Processes – The Disparity between Teacher and Student Knowledge. *Recherches en Didactique des Mathématiques, 11*(1), 65 – 107.

Steinbring, H. (1991c). The Theoretical Nature of Probability in the Classroom. In R. Kapadia & M. Borovcnik (Eds.), *Chance Encounters: Probability in Education* (pp. 135 – 167). Dordrecht: Kluwer Academic Publishers.

Steinbring, H. (1993). Die Konstruktion mathematischen Wissens im Unterricht – Eine epistemologische Methode der Interaktionsanalyse. *Journal für Mathematik-Didaktik, 14*(2), 113 – 145.

Steinbring, H. (1994). Dialogue between Theory and Practice in Mathematics Education. In R. Biehler, R. W. Scholz, R. Sträßer & B. Winkelmann (Eds.), *Didactics of Mathematics as a Scientific Discipline* (pp. 89 – 102). Dordrecht: Kluwer Academic Publishers.

Steinbring, H. (1997). Epistemological Investigation of Classroom Interaction in Elementary Mathematics Teaching. *Educational Studies in Mathematics, 32*(1), 49 – 92.

Steinbring, H. (1998a). Epistemological Constraints of Mathematical Knowledge in Social Learning Settings. In A. Sierpinska & J. Kilpatrick (Eds.), *Mathematics Education as a Research Domain: A search for identity* (pp. 513 – 526). Dordrecht: Kluwer Academic Publishers.

Steinbring, H. (1998b). Elements of Epistemological Knowledge for Mathematics Teachers. *Journal of Mathematics Teacher Education, 1*(2), 157 – 189.

Steinbring, H. (1998c). From »Stoffdidaktik« to Social Interactionism: An Evolution of Approaches to the Study of Language and Communication in German Mathematics Education Research. In H. Steinbring, M. G. Bartolini Bussi & A. Sierpinska (Eds.), *Language and Communication in the Mathematics Classroom* (pp. 102 – 119). Reston, Virginia: National Council of Teachers of Mathematics.

Steinbring, H. (1998d). Mathematical Understanding in Classroom Interaction: The Interrelation of Social and Epistemological Constraints. In F. Seeger, J. Voigt & U. Waschescio (Eds.), *The Culture of the Mathematics Classroom* (pp. 344 – 372). Cambridge, UK: Cambridge University Press.

Steinbring, H. (2000). *Epistemologische und sozial-interaktive Bedingungen der Konstruktion mathematischer Wissensstrukturen (im Unterricht der Grundschule). (Abschlussbericht zu einem DFG – Projekt)*. Dortmund: Universität Dortmund.

Steinbring, H. (2001). Mathematical Symbols: Means of Describing or of Constructing Reality? In C. Bergsten (Ed.), *Conceptions of Mathematics. Proceedings of Norma 01 the Third Nordic Conference on Mathematics Education 2001* (pp. 344 – 372). Norköpping.

Stigler, J. W. & Hiebert, J. (1999). *The Teaching Gap*. New York: The Free Press.

Voigt, J. (1984). *Interaktionsmuster und Routinen im Mathematikunterricht – Theoretische Grundlagen und mikroethnographische Falluntersuchungen.* . Weinheim : Beltz .

Voigt, J. (1991). Die mikroethnographische Erkundung von Mathematikunterricht – Interpretative Methoden der Interaktionsanalyse. In H. Maier & J. Voigt (Eds.), *Interpretative Unterrichtsforschung* (pp. 79 – 116). Köln: Aulis.

Voigt, J. (1994). Negotiation of Mathematical Meaning and Learning Mathematics. *Educational Studies in Mathematics*(26), 275 – 298.

Voigt, J. (1998). The Culture of the Mathematics Classroom: Negotiating the Mathematical Meaning of Empirical Phenomena. In F. Seeger, J.Voigt & U. Waschescio (Eds.), *The Culture of the Mathematics Classroom* (pp. 191 – 220). Cambridge, UK: Cambridge University Press.

Wagner, R. (1981). *The Invention of Culture*. Chicago: University of Chicago Press.

Wagner, R. (1986). *Symbols that Stand for Themselves*. Chicago: University of Chicago Press.

Wenger, E. (1998). *Communities of Practice. Learning, Meaning, and Identity*. Cambridge: Cambridge University Press.

Wilder, R. L. (1981). *Mathematics as a Cultural System*. London: Pergamon Press.

Wilder, R. L. (1986). The Cultural Basis of Mathematics. In T. Tymoczko (Ed.), *New Directions in the Philosophy of Mathematics* (pp. 185 – 199). Boston: Birkhäuser.

Winter, H. (1975). Allgemeine Lernziele für den Mathematikunterricht? *Zentralblatt für Didaktik der Mathematik., 7*(3), 106 – 116.

Winter, H. (1982). Das Gleichheitszeichen im Mathematikunterricht der Primarstufe. *mathematica didactica, 5*, 185 – 211.

Winter, H. (1984). Begriff und Bedeutung des Übens im Mathematikunterricht. *mathematik lehren, 2*, 4 – 16.

Wittmann, E. C. (1984). Teaching Units as the Integrating Core of Mathematics Education. *Educational Studies in Mathematics, 15*, 25 – 36.

Wittmann, E. C. (1995a). Mathematics Education as a „Design Science". *Educational Studies in Mathematics, 29*, 355 – 374.

Wittmann, E. C. (1995b). Aktiv-entdeckendes Lernen im Rechenunterricht – vom Kind und von der Sache aus. In G. N. Müller & E. C. Wittmann (Eds.), *Mit Kindern rechnen. Beiträge zur Reform der Grundschule.* (Vol. 96, pp. 10 – 41). Frankfurt a.M.: Arbeitskreis Grundschule.

Wittmann, E. C. (1997). Das Projekt "mathe 2000" – Modell für fachdidaktische Entwicklungsforschung. In G. N. Müller, H. Steinbring & E. C. Wittmann (Eds.), *Bilanz und Perspektiven. Festschrift zum 10jährigen Bestehen des Projekts "mathe 2000" an der Universität Dortmund* (pp. 41 – 65). Leipzig: Ernst Klett Grundschulverlag.

Wittmann, E. C. (2001). Developing Mathematics Education in a Systemic Process. *Educational Studies in Mathematics, 48*(1), 1 – 20.

Wittmann, E. C. & et al. (1996). *Das Zahlenbuch. Mathematik im dritten Schuljahr. Lehrerband.* Stuttgart: Klett.

Wittmann, E. C. & Müller, G. (1992). *Handbuch produktiver Rechenübungen. Bd. 2: Vom halbschriftlichen zum schriftlichen Rechnen*. Stuttgart: Klett.

Wittmann, E. C. & Müller, G. N. (1990). *Handbuch produktiver Rechenübungen (Bd. 1: Vom Einspluseins zum Einmaleins)*. Stuttgart: Klett.

Wood, T. (1994). Patterns of Interaction and the Culture of Mathematics Classrooms. In S. Lerman (Ed.), *Cultural Perspectives on the Mathematics Classroom* (pp. 149 – 168). Dordrecht: Kluwer Academic Publishers.

Wood, T. (1996). Events in Learning Mathematics: Insights from Research in Classroom. *Educational Studies in Mathematics, 30*(1), 85 – 105.

Wood, T. (1998). Alternative Patterns of Communication in Mathematics Classes: Funneling or Focusing? In H. Steinbring, M. G. Bartolini Bussi & A. Sierpinska (Eds.), *Language and Communication in the Mathematics Classroom* (pp. 167 – 178). Reston, Virginia: National Council of Teachers of Mathematics.

SUBJECT INDEX

INDEX OF NAMES

Mathematics Education Library

Managing Editor: A.J. Bishop, Melbourne, Australia

1. H. Freudenthal: Didactical Phenomenology of Mathematical Structures. 1983
   ISBN 90-277-1535-1 HB; 90-277-2261-7 PB
2. B. Christiansen, A. G. Howson and M. Otte (eds.): Perspectives on Mathematics Education. Papers submitted by Members of the Bacomet Group. 1986
   ISBN 90-277-1929-2 HB; 90-277-2118-1 PB
3. A. Treffers: Three Dimensions. A Model of Goal and Theory Description in Mathematics Instruction TheWiskobas Project. 1987 ISBN 90-277-2165-3
4. S. Mellin-Olsen: The Politics of Mathematics Education. 1987
   ISBN 90-277-2350-8
5. E. Fischbein: Intuition in Science and Mathematics. An Educational Approach. 1987
   ISBN 90-277-2506-3
6. A.J. Bishop: Mathematical Enculturation. A Cultural Perspective on Mathematics Education.1988 ISBN 90-277-2646-9 HB; 1991 0-7923-1270-8 PB
7. E. von Glasersfeld (ed.): Radical Constructivism in Mathematics Education. 1991
   ISBN 0-7923-1257-0
8. L. Streefland: Fractions in Realistic Mathematics Education. A Paradigm of Developmental Research. 1991 ISBN 0-7923-1282-1
9. H. Freudenthal: Revisiting Mathematics Education. China Lectures. 1991
   ISBN 0-7923-1299-6
10. A.J. Bishop, S. Mellin-Olsen and J. van Dormolen (eds.): Mathematical Knowledge: Its Growth Through Teaching. 1991 ISBN 0-7923-1344-5
11. D. Tall (ed.): Advanced Mathematical Thinking. 1991 ISBN 0-7923-1456-5
12. R. Kapadia and M. Borovcnik (eds.): Chance Encounters: Probability in Education. 1991 ISBN 0-7923-1474-3
13. R. Biehler, R.W. Scholz, R. Sträßer and B. Winkelmann (eds.): Didactics of Mathematics as a Scientific Discipline. 1994 ISBN 0-7923-2613-X
14. S. Lerman (ed.): Cultural Perspectives on the Mathematics Classroom. 1994
    ISBN 0-7923-2931-7
15. O. Skovsmose: Towards a Philosophy of Critical Mathematics Education. 1994
    ISBN 0-7923-2932-5
16. H. Mansfield, N.A. Pateman and N. Bednarz (eds.): Mathematics for Tomorrow's Young Children. International Perspectives on Curriculum. 1996
    ISBN 0-7923-3998-3
17. R. Noss and C. Hoyles:Windows on Mathematical Meanings. Learning Cultures and Computers. 1996 ISBN 0-7923-4073-6 HB; 0-7923-4074-4 PB
18. N. Bednarz, C. Kieran and L. Lee (eds.): Approaches to Algebra. Perspectives for Research and Teaching.1996 ISBN 0-7923-4145-7 HB; 0-7923-4168-6, PB
19. G. Brousseau: Theory of Didactical Situations in Mathematics. Didactique des Mathématiques 1970-1990. Edited and translated by N. Balacheff, M. Cooper, R. Sutherland and V. Warfield. 1997 ISBN 0-7923-4526-6
20. T. Brown: Mathematics Education and Language. Interpreting Hermeneutics and Post-Structuralism. 1997 ISBN 0-7923-4554-1 HB
    Second Revised Edition. 2001 ISBN 0-7923-6969-6 PB

Mathematics Education Library

21. D. Coben, J. O'Donoghue and G.E. FitzSimons (eds.): Perspectives on Adults Learning Mathematics. Research and Practice. 2000 ISBN 0-7923-6415-5
22. R. Sutherland, T. Rojano, A. Bell and R. Lins (eds.): Perspectives on School Algebra. 2000 ISBN 0-7923-6462-7
23. J.-L. Dorier (ed.): On the Teaching of Linear Algebra. 2000 ISBN 0-7923-6539-9
24. A. Bessot and J. Ridgway (eds.): Education for Mathematics in the Workplace. 2000 ISBN 0-7923-6663-8
25. D. Clarke (ed.): Perspectives on Practice and Meaning in Mathematics and Science Classrooms. 2001 ISBN 0-7923-6938-6 HB; 0-7923-6939-4 PB
26. J. Adler: Teaching Mathematics in Multilingual Classrooms. 2001 ISBN 0-7923-7079-1 HB; 0-7923-7080-5 PB
27. G. de Abreu, A.J. Bishop and N.C. Presmeg (eds.): Transitions Between Contexts of Mathematical Practices. 2001 ISBN 0-7923-7185-2
28. G.E. FitzSimons:What Counts as Mathematics? Technologies of Power in Adult and Vocational Education. 2002 ISBN 1-4020-0668-3
29. H.Alrø andO. Skovsmose:Dialogue and Learning in Mathematics Education. Intention, Reflection,Critique.2002 ISBN 1-4020-0998-4 HB;1-4020-1927-0 PB
30. K. Gravemeijer, R. Lehrer, B. van Oers and L. Verschaffel (eds.): Symbolizing, Modeling and Tool Use in Mathematics Education. 2002 ISBN 1-4020-1032-X
31. G.C. Leder, E. Pehkonen and G. Tőrner (eds.): Beliefs: A Hidden Variable in Mathematics Education? 2002 ISBN 1-4020-1057-5 HB; 1-4020-1058-3 PB
32. R. Vithal: In Search of a Pedagogy of Conflict and Dialogue for Mathematics Education. 2003 ISBN 1-4020-1504-6
33. H.W. Heymann: Why Teach Mathematics? A Focus on General Education. 2003 Translated by Thomas LaPresti ISBN 1-4020-1786-3
34. L. Burton: Mathematicians as Enquirers: Learning about Learning Mathematics. 2004 ISBN 1-4020-7853-6 HB; 1-4020-7859-5 PB
35. P. Valero, R. Zevenbergen (eds.): Researching the Socio-Political Dimensions of Mathematics Education: Issues of Power in Theory and Methodology. 2004 ISBN 1-4020-7906-0
36. D. Guin, K. Ruthven, L. Trouche (eds.) The Didactical Challenge of Symbolic Calculators: Turning a Computational Device into a Mathematical Instrument 2005 ISBN 0-387-23158-7
37. J. Kilpatrick, C. Hoyles, O. Skovsmose (eds. in collaboration with Paola Valero) Meaning in Mathematics Education. 2005 ISBN 0-387-24039-X
38. H. Steinbring. The Construction of New Mathematical Knowledge in Classroom Interaction: An Epistemological Perspective. 2005. ISBN 0-387-24251-1